NSF MOSAIC READER

ECOLOGY
IMPACTS AND IMPLICATIONS

NSF MOSAIC READER

ECOLOGY
IMPACTS AND IMPLICATIONS

AVERY PUBLISHING GROUP INC.
Wayne, New Jersey

The articles contained in this volume were selected from original works that appeared in *Mosaic* magazine. They are reprinted by permission of the National Science Foundation.

Mosaic is published six times yearly as a source of information for the scientific and educational communities served by the National Science Foundation. For more information regarding *Mosaic,* please direct your inquiries to: Editor, *Mosaic,* National Science Foundation, Washington, D.C. 20550.

The publisher is indebted to Warren Kornberg, Editor of *Mosaic,* for his editorial guidance, his invaluable suggestions, and his patience. Avery also wishes to thank the members of its own editorial board for their help in article selection. Our thanks go to Paul Biersuck, Department of Biology, Nassau Community College; Mel Gorelick, Department of Biological Sciences, Queensborough Community College; John Burkart and Loretta Chiarenza, Department of Biology, State University of New York at Farmingdale; Bernard Tunik, Department of Biology, State University of New York at Stony Brook; John Maiello, Department of Biology, Rutgers— The State University of New Jersey; and Donald Wetherell, Biological Science Group, University of Connecticut at Storrs.

Cover design by Martin Hochberg.
Cover photo credit: Martin Hochberg.
In-house editor: Joanne Abrams.

Copyright © 1983 by Avery Publishing Group, Inc.

ISBN 0-89529-176-2

All rights reserved. No part of this publication may be reproduced, stored in a retrieval system, or transmitted in any form or by any means, electronic, mechanical, photocopying, recording or otherwise, without the prior written permission of the copyright owner.

Printed in the United States of America

10 9 8 7 6 5 4 3 2

CONTENTS

Introduction 1

Crossroads For Tropical Biology 2

    Tropical forests are home to two thirds of the world's plant and animal species. Many of these life forms have not yet been identified and catalogued, much less studied for their contribution to the ecosystem or their possible use as sources of food or medicine. Nonetheless, tropical forests are rapidly disappearing—falling before the pressure of population and the onrush of development. Developmental efforts are doomed to failure, however, if they neglect to take into account the unique structure of tropical forest ecosystems; applying techniques learned in the temperate zones will simply not do. This reading examines the attempts being made to study and, where possible, preserve tropical rain forests in the face of heedless commercial exploitation and other destructive forces.

Antarctica: No Catch Limit Yet 11

    As the nations of the world expand their search for resources, the antarctic, as one of the few remaining unexploited frontiers, receives increasing attention. This is true not only because of our constant quest for minerals, in which Antarctica and the waters around it are rich, but also because of the world's increasing need for food. This reading examines what is known—and what is unknown—about critical antarctic oceanic species, and the current efforts to prevent runaway exploitation by first learning how the pieces of the antarctic ecosystem fit together and why balances must be maintained if resources are to be protected.

Probing the Bering Sea Shelf 19

    This reading highlights the ecology of the Bering Sea shelf, an area whose uniquely structured ecosystem supports an enormous number of high-order consumers, including many birds, seals, and fish. Thus the Bering Sea shelf fails to conform to the accepted model of marine ecosystems—a model that places mathematical limits on the proportions of organisms that may exist at each level of the oceanic pyramid of life. This intriguing fact has inspired a number of studies, the most significant of which is PROBES (Processes and Resources of the Bering Sea Shelf). The author examines early data gathered by PROBES, and looks ahead to the results of an ocean ecosystem study that is both broad and intensive.

Acid From the Sky 27

Acid precipitation—that is, rain and snow acidified by, among other agents, commercially emitted compounds—has long been recognized as an environmental threat. Ecologists know that acid-polluted precipitation has already rendered many lakes incapable of supporting life, and that "acid rain" may stunt forest growth and reduce agricultural productivity. At this time, the mechanisms of this increasingly common phenomenon are little understood; indeed, scientists are often unable to confidently identify the source of any one acid precipitation episode or to explain the earth's accomodation to naturally occurring acid precipitation. Intensive studies of the problem are being conducted both here and abroad in an effort to bring a growing problem under control. The author discusses the many aspects of acid precipitation and reviews some of the more significant research efforts currently underway.

The Ocean in a Test Tube 33

In the early 1970's, it was a generally accepted notion that the oceans were dying—that pollution was killing everything off. While this concept, popularized by explorers such as Jacques Cousteau and Thor Hyerdahl, has since been refuted by studies that reveal clean and healthy open ocean environments, researchers are nonetheless concerned about the effects of gradual toxic buildups on the ocean's ecosystems. In an effort to prevent the threat of the 70s from becoming a reality, sophisticated techniques have been developed to detect such slow accumulations and to anticipate their effect on marine life. This reading explores the intriguing tools being used to compile the necessary data, and discusses the surprising conclusions that have emanated from multinational studies.

Mount St. Helens: The Ecology of a Holocaust 40

When Mount St. Helens errupted on May 18, 1980, some 600 square kilometers of wilderness were abruptly transformed into an awesome wasteland of dried mud, felled trees, and ash. Although full recovery of the area will take more than a century, the reestablishment of life began almost immediately, providing scientists with a unique opportunity to witness the pattern of species renewal and to test theories of ecological equilibrium. Even before the ash was cool, teams of researchers gathered to observe animal, vegetable, and even bacterial life forms striving to survive or reestablish themselves. This article examines the early stages of biological recovery and discusses the additions to human knowledge that can be derived even from catastrophes of such magnitude as this.

Glossary **47**
Index **51**

INTRODUCTION

Mosaic, the source of articles in this reader, is the bimonthly magazine of the National Science Foundation. Its purpose is to keep nonspecialists in any of the sciences aware of the ferment at the frontiers of many scientific research disciplines and the research trends out of which tomorrow's scientific and engineering advances will emerge.

Mosaic's purpose is to explore the thinking of researchers about both the current and future status of their science. Its articles provide insight into the problems facing investigators in virtually every research area, and explore the ways they seek to overcome those problems—often by crossing traditional disciplinary lines.

Prepared by experienced science journalists and authenticated by scientists, these articles reveal the processes of science, as well as its progress. They not only report on day-to-day advances but offer perspectives on what science is.

For the *Mosaic Reader Series,* recent issues of the magazine have been surveyed, and groups of articles assembled to provide broad, pertinent overviews of segments of scientific research. This reader on ecology and its implications focuses on the extent to which human beings are a component of their environment and must consider their interaction with it. Some forces that shape the context in which living things develop are beyond us; natural phenomena such as major volcanoes affect the environment as they provide unparalleled opportunities to study it. But there are artificial phenomena too—things human beings do to their world that must also be understood and often modified or prevented.

To make sense of any of this, there must be a body of baseline knowledge about the functioning of ecological systems. This foundation is beginning to be laid. If anything comes out of this reader, it should be an understanding of the complexity of interacting forces and life forms. Ecologists must try to come to terms with this complexity as they learn what must be done to preserve our environment for—and from—ourselves.

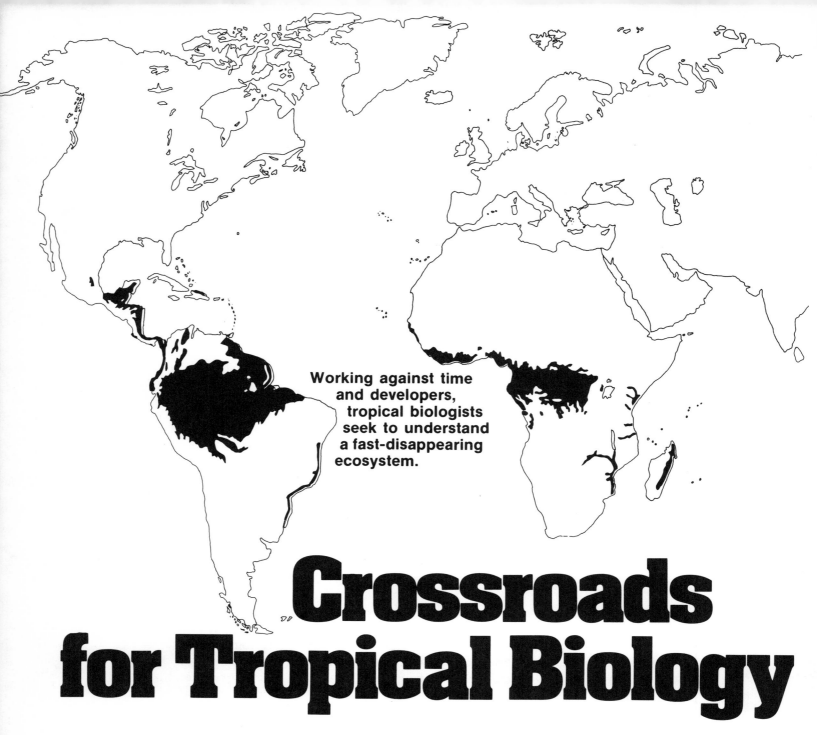

Working against time and developers, tropical biologists seek to understand a fast-disappearing ecosystem.

Crossroads for Tropical Biology

Two and a half million species of plants and animals are in danger of becoming extinct before they become known. This is by computation, not count. But that such guesswork is necessary underlines the concern of biologists worldwide: If something is not done to curtail the destruction of tropical forest ecosystems, the webs of life that define them and limit or prescribe what can be produced where they stand will never be known. The concern is far from being a case of research *versus* human need. Tropical ecosystems are sufficiently unique and so little understood that their unstudied destruction threatens to forestall solution to those human needs—food, energy and economic development—that are motivating the onslaughts in the first place.

"We estimate that more than 20 hectares of tropical forest are cleared each minute; that comes to almost 11 million hectares a year," says Peter H. Raven, chairman of the National Academy of Sciences' Committee on Research Priorities in Tropical Biology. "Out of 16 million square kilometers of forests that once girdled the earth between the Tropics of Cancer and Capricorn, only about 9 million remain. The destruction of the remaining forests before an onslaught of food- and energy-short people, whose numbers will double by the year 2000, may be virtually complete in 50 years if something is not done." Further, says Raven, "Two-thirds of the world's 4.5 million plant and animal species live only in the tropics; species are becoming extinct on a scale that has never been equaled, even by the disappearance of the dinosaurs."

Other tropical biologists do not consider this an exaggeration. A comparison of known species leads them to conclude

Humid forests. The lush denseness of humid forest ecosystems is rapidly disappearing as exploitation outpaces research and conservation around the world.

New York Botanical Garden/Ghillean Prance

that twice as many varieties of life exist in the tropics as in the world's temperate regions. Catalogues of temperate-zone species list about a million plants and animals, and it is estimated that half a million more await detection. Tropical catalogues, then, should contain three million names. But the catalogued total is only 500,000.

Some biologists believe that 4.5 million is a highly conservative estimate of the number of species on earth, and that the varieties of "missing" life in the tropics, particularly insects and lower plants, far exceed 2.5 million.

Slash and burn. Clearing tropical forest land by the classical slash-and-burn technique destroys the living network that keeps the nutrient-poor tropical soil fertile.

University of Georgia/Carl Jordan

Reason to learn

Extinction of millions of undiscovered species would be much more than an appalling loss of knowledge and understanding, says Raven, who is director of the Missouri Botanical Garden in St. Louis. There are also compelling economic reasons for curbing the destruction.

"For one thing," Raven declares, "an understanding of the way in which tropical forest ecosystems function presumably holds the key to development of stable agricultural systems. Most rain forests are in less-developed countries where population increases and food shortages are most critical. For another thing, 40 percent of all prescriptions written in the United States today contain at least one ingredient originally made from a plant. Not one in a hundred of the tropical plants in danger of becoming extinct has been examined for useful drugs or other chemicals such as insecticides. In addition, development of important new tropical crops such as rubber and palm oil during the past 150 years should motivate us not to lose other possibilities before they have been examined by a single scientist."

The task of the NAS committee, as its name implies, is to recommend action that will yield the most relevant data about tropical forests before they disappear. The committee defines relevance in both basic and practical terms. It not only wants scientists to learn the structure and function of rain forests, it wants the information to be useful in stemming destruction and providing social and economic benefits.

The forests that concern the NAS committee and other biologists lie in three principal areas—Latin America, Africa and Indonesia/Malaysia. They are warm, wet, verdant places that laymen usually refer to as "jungles." Biologists scorn this word in favor of "tropical rain forests" or "humid tropical forests." In these forests, temperatures rarely drop below 20 degrees centigrade and rainfall exceeds evaporation. At least 200 centimeters of rain falls on them annually, though some locations may get as much as 1,500.

The American rain forest, centered on the Amazon River basin and extending north into Mexico, is the most extensive. Its 4.2 million square kilometers represent about a sixth of the broad-leafed forest regions of the world. Raven estimates that at least 400,000 and perhaps as many as a million species live in the Amazon Basin alone.

Halfway around the world, the Indo-Malaysian or Asian rain forest covers about 2.6 million square kilometers. It is the richest in plant and animal varieties, including many species found nowhere else. The African rain forest, centered on the Congo Basin, covers about 2.0 million square kilometers and is considered the poorest in species number. Smaller rain forests occur in India, Sri Lanka, Madagascar and other Indian Ocean islands.

The NAS committee estimates that the American rain forest originally covered a third more area before it was cut back by timber, farming and ranching operations. Two-thirds of the Asian forests have been axed and burned away, mainly by timber companies. "Most of the remaining high rain forest in Southeast Asia is under logging license," the International Union for Conservation of Nature and Natural Resources reports, "and timber resources are unlikely to last beyond the end of the century."

Half of the African rain forest is gone, as well as a staggering two-thirds of the tropical forests in India, Sri Lanka and Burma. Raven says that, given the present rates of utilization, in 20 years humid forests in Madagascar and the Philippine Islands will no longer exist. The situation in Madagascar particularly troubles tropical biologists; its rain forest holds a unique biota which evolved in long isolation.

Researchers believe most of the unknown species in these regions are insects and other arthropods. Plants come next. A botanist about to leave a desk job for field work in New Guinea expects to find dozens of new species of plants in the coffee family, including some of economic value. But higher animal species continue to be discovered in rain forests as well. A new warbler was identified in the El Yunque forest of Puerto Rico less than 20 years ago. Undergraduate students working on an NSF-sponsored summer research program described a new bird genus endemic to Hawaii in 1974. Other investigators found a new genus of peccary in Paraguay in 1975.

The NAS committee, seeking to shape the spearhead of a major research effort, will recommend a program of collecting in endangered areas. "We want to combine collecting with a study of how the species function in the ecosystem and with a chemical screening of plants to assess economic potential," Raven states. A number of regional inventories already are under way. The Missouri Botanical Garden, for example, is completing a 26-year survey of the flora of Panama. At the same time, Missouri Botanical Garden scientists are attempting to catalogue all the plants in Nicaragua and, in collaboration with Chicago's Field Museum, in Peru.

Projeto Flora

The most comprehensive effort to inventory rain forest species is under way in Brazil. The project there, called Projeto Flora Amazônica, is a joint U.S.-Brazilian program representing a major attempt to collect plants in rain forest areas being threatened by development.

The United States became involved in Projeto Flora at the invitation of Brazil's Conselho Nacional de Pesquisas (National Research Council); three expeditions to the Amazonian rain forest, coordinated by Ghillean T. Prance of the New York Botanical Garden, took place in 1977 and 1978 and six more are planned for 1979 and 1980.

Prance and his Brazilian and North American colleagues "collected more new

species on these trips than on any expedition in which I ever participated," Prance reports. "The largest number came from an endangered area along a new north-south highway linking Santarém on the Amazon River with Cuiabá, 1,400 kilometers to the south. We are screening our collection from there and from other locations, looking for new insecticides, food crop varieties and drugs."

Several plants from the Amazon have yielded insect repellents, says Prance, and there is special interest in tropical species as a potential source of drugs for difficult-to-cure diseases. "Recently discovered compounds effective against leukemia come from the tropical periwinkle, which grows mainly in Madagascar,"

Declining yield. Manioc, seen behind each of these Venezuelan farmers, produces a good crop on slashed- and burned-over soil in the first year (above) and a skimpier crop (below) in the second.
University of Georgia/Carl Jordan

Prance reports, and "it is not far fetched to believe that we might discover other drugs effective against cancer."

Amazon Indians introduced Prance and his colleagues to the beka vine (*Curarea tecunarum*), the bark of which is used by Indian women in a contraceptive preparation. Prance turned the plant over to a testing program, sponsored by the World Health Organization, at the University of Chicago. So far, test results are positive; components of the bark prevent laboratory rats from becoming pregnant.

Projeto Flora expeditions also collected along part of the Tocantins River, which will be dammed at Tucuruì in 1982 to create a new lake 175 kilometers long. The dam will provide hydroelectric power for the city of Belém and for an iron mine in the Carajás Mountains.

"Hydroelectric power in that area is a good balance between conservation and development," Prance observes. "It is pollution-free and, placed in a location with few or no endemic species, it does minimum damage to the ecosystem."

For other areas in Brazil, the match between development and the ability of the ecosystem to sustain it may be less fortunate. One U.S.-owned, international conglomerate has cut heavily into its 400,000-hectare (million-acre) Amazon forest holdings, replacing native species with plantations of fast-growing Caribbean pine and gmelina trees. At the southern edge of the rain forest, a European concern has killed 200,000 hectares of trees by chemical spraying and burning, planting grass in the ashes in the hope of raising cattle for the low-grade beef/hamburger franchise market around the world.

Had the initiators of these and similar operations consulted tropical biologists in advance, they might have had second thoughts. Cutting down the trees destroys most of the nutrient-conserving mechanisms of a tropical forest ecosystem. Rainwater carries off vital nutrients; the soil becomes sterile, often in two or three years.

Current commercial enterprises, biologists agree, are likely to fall far short of expectations, besides doing considerable, if not irreparable, damage. Exploitation of humid tropical ecosystems is possible, they say, but it must take the characteristics of the ecosystems into account. The knowledge base for such system-sensitive exploitation is among the goals of the joint U.S.-Brazilian Projeto Flora Amazônica.

"On Projeto Flora, our objectives are not confined to collecting plants, storing them in museums and writing papers about them," Prance notes. "We attempt to get information useful for planning both conservation and development, and to make this information available for the social and economic benefit of all concerned."

The Brazilian Government is not unreceptive to conservation efforts, Prance observes. "A new, high-level department has been set up under a Secretary of the Environment, and preserves established by this secretariat have saved a number of species from disappearance. The Brazilian Forest Service plans to create large national parks in the Amazon. A 400,000-hectare park in the Tapajós River basin is under way; others exist on paper."

The San Carlos project

Training and research also are under way in Venezuela. There, U.S., West German and Venezuelan scientists cooperate on the so-called San Carlos project, designed to solve a perplexing contradiction posed by humid tropical ecosystems: their resistance to cultivation despite their apparent biological richness.

Trees so dense that they shut out the sun, the bewildering diversity of life forms, a year-round growing season and the abundance of rain give the impression that these should be among one of the most fruitful regions on earth. Indeed, planners and policy-makers once looked to "jungles" for a solution to the world's food problems. When farms replace the trees, however, yields decline dramatically and the land is abandoned, often by the end of the third year. Why? The answer is in the nature of the system.

The San Carlos project is directed by Ernesto Medina of the Instituto Venezolano de Investigaciones Científicas (Venezuelan Institute for Scientific Research) in Caracas. The Institute of Ecology at the University of Georgia coordinates U.S. participation. The German effort involves the Max Planck Institute and the World Institute of Forestry. Support comes from a variety of sources, including the Venezuelan Science Foundation; Deutsche Forschungsgemeneinschaft, Organization of American States, UNESCO and the National Science Foundation.

Not many years ago, says Carl Jordan of the University of Georgia, the high potential of tropical forests for agricultural productivity was part of the conventional wisdom. But when Jordan and his colleagues looked, what they found was poor soil and productivity no greater than that of temperate forests.

"If you cut down all the vegetation on a plot of rain forest and one in the woodlands of the Eastern United States," says Jordan, "the tropical field fills in with secondary growth much more rapidly. It may take a woodland 20 years to look as luxuriant as a tropical forest looks in two years.

"But when you compare the production of wood in mature forests, you discover that tropical forests are no more productive than a typical forest in the Southern United States." In the tropics, there is a denser growth of leaves than in temperate regions. But wood production is similar.

The 365-day growing season creates another popular fallacy. Frank Golley of the University of Georgia calls it "an advantage for insect pests and for parasites but not for trees." Nor is the abundant rainfall characteristic of tropical forests an unmixed blessing. It leaches nutrients out of the soil; when the soil is coarse and of low fertility, this leaching severely inhibits the establishment of ground cover to protect the soil from further damage by erosion.

This, of course, raises the question: How can poor soil support lush greenery and a dazzling variety of trees, some of which grow as tall as 60 meters? The San Carlos research team counted 200 different species of trees on just four hectares of Venezuelan forestland.

The first explanation of this apparent contradiction came in the late nineteen-sixties, as an aftermath of an Amazon expedition conducted by Nellie Stark of the University of Montana and Frits Went of the University of Nevada. Soil samples they collected contained roots with surprisingly large quantities of a filamentous fungus known as *mycorrhiza*. Individual filaments are finer than a spider web strand, but they form a bridge between fallen leaves and fine roots covering the forest floor.

Nutrient cycling

Stark and Went studied the *mycorrhizae* and, in 1971, hypothesized that the fungi take up nutrients from decomposing organic matter and transfer them directly to the roots of the trees. Thus, nutrients are kept from the poor soil where they could be leached away.

This perception was tested by Jordan, Stark and their colleagues, working near the village of San Carlos in the north-central part of Amazonia. They sprinkled the mat of roots, leaves and fungi lying atop the soil with a radioactive solution of phosphorous and calcium and traced the movement of the radioactive nutrients through the fungi to the roots. "This proved that *mycorrhizae* play a role in the survival of the tropical forest," Jordan comments. "They help answer the question of how poor soil can support such a luxuriant growth."

Soil's bank. The shallow surface-mat of roots, leaves and fungi (above) is the nutrient bank in tropical soils even if soils are 40 centimeters deep. Filamentous fungi (below) enshroud roots and transfer nutrients directly back to rain-forest trees.

University of Georgia/Carl Jordan

When researchers studied the root mat and trees closely, they found that fungi are not the sole, or even the most important, mechanism by which nutrients are conserved and recycled. Rain falling through the crowns of trees picks up nutrients. When it reaches the ground, the nutrients are absorbed directly onto the root mat and humus. "This appears to be the major nutrient-conserving mechanism in tropical rain forests," says Jordan. "It is this mechanism that is destroyed by cutting down the trees."

Other nutrient-conserving mechanisms involve microfauna such as bacteria, algae and lichens which decompose leaves and twigs, taking up nutrients and keeping them cycling within the root mat. Algae and lichens living on the bark and attached leaves also fix nitrogen, which may be utilized by the tree.

"Further, in forests where nutrients are more plentiful," says Jordan, "trees obtain enough energy to afford 'throwaway' leaves. In nutrient-poor forests, a better strategy is to stay evergreen and save the energy that would go into making new leaves each year." Rain-forest trees on nutrient-poor soils, consequently, possess thick, heavy, waxy leaves, efficient for retaining water and nutrients which often move back into the tree before the leaves fall.

The leaves are also relatively insect-proof. "On the ground," Golley notes, "one sees intact leaves, rather than leaves riddled and scalloped by insect bites as is common in U.S. woodlands." The trees invest relatively large amounts of energy in the production of such secondary compounds as alkaloids and phenols, which make the leaves toxic or repellent to insects.

Preserving the capital

Without the nutrient-conserving mechanisms of the undisturbed forest, rains rapidly leach away nutrients. The peasant farmer knows his crop yield will drop drastically the second or third year after planting; the next year, he abandons his plot and moves on to a new site. But there is another adaptation made by local farmers that might have been a sign to loggers and ranchers: They work only small plots.

"Local people clear less than a hectare when they practice slash-and-burn agriculture," Jordan points out. "The abandoned plot is small enough so that seeds and organic matter from the surrounding forest easily move in and fill it up. Eventually the land can be farmed again. However, when tracts the size of Connecticut are cleared of trees, the forest is in danger of becoming a biological desert."

In Brazil, the U.S. forestry firm's first crop of pulpwood trees is said to be coming along well. But the concern is rapidly using up the area's nutrient capital, according to Jordan. Many tropical biologists doubt it will get a second crop after workers haul out the first trees. And in Costa Rica, Brazil and other places where the forest has been converted to pasture, nutrient loss is accelerated by the cattle themselves, compacting the soil with their hooves. Nutrients run off the land "like rain off a paved road," says Jordan. Unless a new root mat forms with associated *mycorrhizae*, this land will also become sterile.

To extend this insight to the ecosystem, the San Carlos team marked off plots near the town of San Carlos on the Rio Negro, close to the borders of Brazil and Colombia. For a year they studied nutrient cycling on the undisturbed plots. Then, under the team's direction, local people cleared half of the plots and planted manioc, just as they would on their own land. North Americans eat manioc only as tapioca, but this starch forms a basic staple in the diet of many Latin Americans. The German group working on the project planted rubber trees nearby in sandy soil that becomes flooded during part of the year.

"We got a good manioc crop the first year, as expected," Jordan says. "The second year, yield dropped off markedly. Next year the plots will not be worth farming, but we will have obtained the first accurate measurements of changes in forest chemistry and productivity resulting from slash-and-burn agriculture."

The German experiment had a different consequence. Though the rubber trees grew only six feet in three years, Jordan notes, forest removal and replanting "has not turned the forest into a sterile desert."

As Jordan and Golley see the inferences of this work, no biologically sound way exists to convert rain forests to farms, pastures or plantations on a large scale. What may be feasible, Golley says, "is a mixed crop: root crops like manioc, ground covers, legumes and small trees. These might be planted in a mosaic, in such a way that the root mat would not be disturbed and plants always would be available to cycle nutrients." It also should be possible to plant, in ways that do not destroy the primary forest structure, oil palm, rubber, banana and cashew trees.

Tropical reclamation

John J. Ewel of the University of Florida in Gainesville agrees. He is trying to find out how to utilize rain forest areas by studying the secondary growth that covers scars left by slash-and-burn

Collecting, recording. U.S. and Brazilian botanists collect plants in threatened areas of the Brazilian Amazon, which is high in endemic species.

New York Botanical Garden/Ghillean Prance

agriculture. "Natural secondary growth conserves available nutrients and reestablishes mechanisms to cycle them," Ewel observes. "It has the kinds of structural and functional properties that we should design into ecosystems modified by man to satisfy food and fiber needs."

Ewel works with 24 plots of 200 square meters each, located near Turrialba, Costa Rica. He uses four types of plot: one resembling local agricultural fields, a second containing secondary vegetation invading a typical slash-and-burn clearing, a third in which Ewel builds ecosystems "from scratch," using species similar to those that grow naturally in that area, and a fourth in which he and his colleagues attempt to increase species richness by adding cuttings and seeds to the natural regrowth.

In this work, Ewel applies knowledge gleaned from previous work on the appetites of insects that feed on rain forest plants. When a single, highly edible species grows by itself in a clearing, the predators consume it at a high rate. But when it is interspersed with a variety of other plants, its survival improves markedly; the desirable plant becomes harder to find, and insects settle for other things to eat. Farmers, therefore, should mix tasty crops with plants less desirable from an insect's point of view.

On the experimental plots, Ewel wants to prevent the wiping out of nutrients once a crop is harvested. To do this, among other things, he'll try double-cropping—planting a second crop before harvesting the first. In February 1979, Ewel cleared the trees and began planting. He started his man-made ecosystem with short-lived leafy plants similar to those that naturally colonize clearings. "These plants don't 'hook up' to the *mycorrhizae*," Ewel points out. "They use all their energy to grow and spread as rapidly as possible."

Natural pioneer plants possess toxins to repel insects; most food crops have had these defenses bred out of them. But Ewel is not trying to obtain a profitable yield—only to keep nutrients in the soil. If he proves that this can be done with food plants as well as natural vegetation, farmers can grow the food crops and use insecticides to prevent the crops from being eaten.

"Before we harvest the short-lived species, we'll plant intermediate crops including cassava (manioc), banana trees and grains such as corn, rice and sorghum. The fungi quickly attach to these plants. Finally, we'll put in long-lived crops such as coffee and cacao. In nature, the secondary growth becomes progressively larger, woodier and more complex. Complexity both reduces insect predation on a single species and produces more root systems to increase nutrient retention."

Building on the know-how

More experimental sites such as those set up at Turrialba and San Carlos are likely to be recommended by the NAS committee. "We have been considering the idea of selecting particular sites for intensive study in different parts of the tropics," notes Raven. "The committee does not advocate establishment of new stations and projects run by U.S. scientists. Any recommendation along these lines would take advantage of existing research centers in various countries and would involve joint participation with local researchers."

In some tropical regions, selected areas have been permanently preserved. Costa Rica has taken the lead in this with more hectares in national parks in proportion to its area than any other tropical country. "In the past eight years Costa Rica has established an outstanding park network and park service," notes Ewel. "A training program exists for park rangers, and two national conservation groups are active in public education programs." In Venezuela, "the entire Amazon territory enjoys a type of park status," Golley points out. "Foreign logging and other commercial ventures are strongly discouraged. Pressure to develop in the rain forests is eased by revenues and jobs created by the oil industry."

In Brazil, large preserves exist on paper, and conservation awareness continues to grow. The possibility of large preserves to protect concentrations of endemic species appears bleakest in Africa and Asia.

Raven's National Academy committee probably will recommend research to determine the optimum size of rain forest preserves. Some of these might be international in scope. Large animals such as jaguars and tapirs are an integral part of the ecosystem and need plenty of room to hunt. "It has been found that Asian hornbills, which live in the tops of trees, need up to 10,000 square kilometers [for a genetically viable population] to survive," notes Ewel. "Preserves likely will be organized on a national basis, and this may be too much area for small countries in Asia to give up."

"We need to preserve not just individual species, but relationships that have evolved over millions of years," Prance adds. "Brazil nut trees in the Amazon require a certain species of large bee for pollination. Other trees depend on particular wasps, moths, humming birds, bats and other insects and birds for fertilization. The preserve size must take into account the range of these animals. Everything in the ecosystem is interwoven, and we must preserve all the relationships for tropical forests to be self-perpetuating."

A race

Saving tropical forests comes down to a race between commercial operations interested in short-term exploitation and scientists trying to obtain a knowledge base for utilizing these unique ecosystems in a way that will take advantage of their properties. In many places commerce has the advantage at present and immediate utilization is the rule.

Scientists still struggle with planning a unified strategy for their part of the race. Every year that slips by another one or two percent of the world's rain forests is lost to development. Raven and Prance want to use data from aircraft and satellites to assess destruction rates and monitor forests on a continuing basis. "Much of the satellite imagery needed for a baseline assessment appears to be available," Prance believes. "What we require is people to interpret the photos and other data."

Besides relevant scientific research, sweeping changes in land-use schemes and other such policy decisions are needed, researchers agree. Ewel points out that between 250 million and 400 million people now live by slash-and-burn agriculture in tropical forests. But if they are part of the problem, it is not a matter of choice; they simply lack alternatives.

"Contrary to popular belief," Ewel says, "they are not a primary cause of deforestation. This is hard, dirty work that no one relishes. That most of these people are forced into it is a reflection of more fundamental causes of their (and our) dilemma: overcrowding on lands suitable for sustained agriculture and unavailability of land-use systems appropriate for sustained productivity in humid tropical forests."

Ewel recommends that agencies of the U.S. and other governments respond with development of new land-use schemes for these forests, increased utilization of less-fragile tropical environments, the screening of technology transfers to ensure that they do not add to the problems and education, not of the rural poor in less-developed nations but of the urban rich who are the decision-makers and investors. He and others want import quotas placed on low-grade beef raised on cleared forests to supply U.S. hamburger chains. Both Ewel and Prance recognize that selective logging is more expensive than clear-cutting, but it is less destructive. They believe schemes could be worked out that would permit logging yet preserve the structure of tropical forests.

The consequences of failure to discover alternatives to destructive development of tropical forest regions appear awesome. The richest source of biological information on earth could be destroyed and plants of potential economic and medical value could be lost, according to an NAS assessment. Also, NAS scientists fear, the course of evolution could be altered. Further, an increase in atmospheric carbon dioxide as a consequence of major destruction of forests could have adverse effects on world climate. There are also shorter term economic consequences.

"Destroying, chipping up and burning forests as a ready source of energy will result in a loss of economic potential for areas now occupied by forests and probably a concomitant setback for the stability of the countries concerned," states the NAS assessment. Adds Raven: "The nations of the world are as interwoven and dependent on each other as the species in a tropical ecosystem. A loss of stability in less-developed countries will affect the stability of the rest of the world."

One scientist compares rain forests to "a sack of uncut diamonds with humankind as the lapidist. Instead of studying and evaluating this resource with the goal of gaining maximum sustained benefit from it, we are throwing the diamonds into a furnace to obtain a short-term source of energy. The consequences may well prove disastrous for humankind as a whole." ●

The National Science Foundation supports research discussed in this article through its Division of Environmental Biology and Division of International Programs.

Antarctica: No Catch Limit Yet

Biologists are concerned that exploitation of antarctic marine resources can have catastrophic effects, unless both resources and effects are better understood.

In the icy waters of the southern oceans, there is a virtually untapped marine protein resource so plentiful that some fishery experts predict that 100 million tons of it could be taken annually without depleting its stocks. This means that the catch of just one zooplankton species—*Euphausia superba*, better known as antarctic krill—could more than double the total annual world fish and shellfish catch, now some 60 to 70 million metric tons a year. Although it is impossible to estimate sustainable yield with any accuracy—some biologists feel that an initial safe annual limit for krill harvesting might not be more than five million metric tons—the enthusiasm for taking krill is real. Several nations have begun exploration, not only of the harvesting possibilities but of the on-site processing and the marketing possibilities as well.

There is some irony in the fact that so much activity—and current or contemplated investment—is based on so few hard data about the ecosystem. Estimates of components of the antarctic biomass, including mammals, birds and fish, are customarily based on what is known of their direct and indirect dependence on krill as a food source. For instance, the widely accepted assumption that there are standing stocks of krill, estimated to be, on the low side, some 50 to 100 million metric tons, can be linked to estimates of populations of some 15 or more million krill-eating crabeater seals and more than 12 million Adélie penguins. Nevertheless, there are so many fundamental things, even about the krill, that are either not yet known or still highly controversial, including not only krill numbers but also their life span, breeding patterns and migration habits, that all such estimates are due for considerable refinement as research proceeds.

The krill feed on smaller zooplankters and plant (phyto) plankters as well. Estimates of the phytoplankton biomass have been put as high as six to ten billion metric tons—a figure to support almost any krill estimate, but which one incredulous marine biologist has calculated to be some 200 grams of phytoplankton per square meter of southern ocean surface.

There is no serious doubt that the Antarctic's waters are highly productive (see "Short food chain," accompanying this article). It is just how productive that has yet to be settled.

Estimates of krill population range widely—if not wildly—from lows between 50 million and 100 million metric tons to highs between 5 and 7 billion metric tons. If, as has been proposed, for instance, crabeater seals consume over 70 million metric tons of krill a year, this alone would be considerably more than the total low estimate for the entire standing stock.

Adding to the confusion and lack of basic knowledge, the geographical boundary of the marine ecosystem, as defined by the Antarctic Convergence, is a changing one. Often drawn as a line, though it isn't really, the Convergence is the shifting front at which the antarctic surface water sinks beneath the less-dense, sub-antarctic surface water. South of the Convergence there is a marked increase in nutrient salt concentrations, and the waters are much richer and more productive than the waters just to the north. The Convergence, however, is not stationary; it fluctuates with changing temperature, pack ice and currents, moving between 54 degrees and 62 degrees south latitude.

The extent of the pack ice south of the Convergence also undergoes seasonal fluctuations, shrinking from 22 million square kilometers in September—the end of winter in the Southern Hemisphere—to 4 million square kilometers in February, in the austral summer (see "Weather from the Ends of the Earth" and "Tales the Ice Can Tell," in this *Mosaic*). Pack ice generally contributes to a hospitable environment for marine life; it almost totally eliminates wave action and turbulence.

The gaps in knowledge of biological marine resources in the Antarctic are

---

Antarctic ecosystem. Major factors underlying the biomass concentrations in the waters around Antarctica include the krill population, water depth, ice extent and the location of the broad, shifting belt called the Antarctic Convergence.

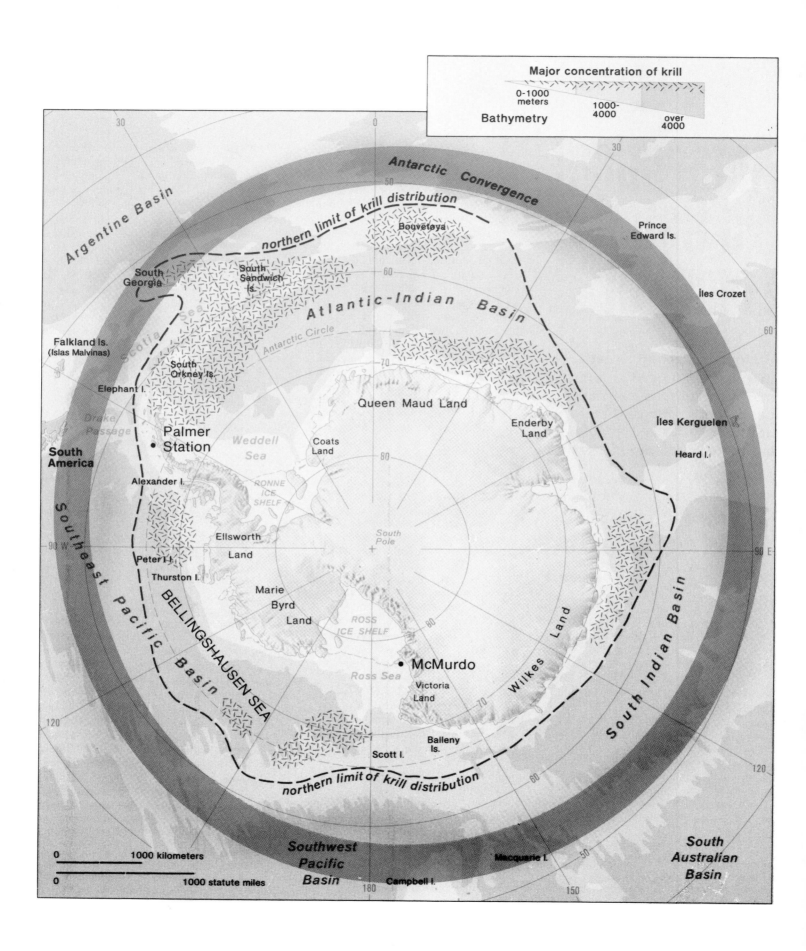

partly explained by the absence so far of a concerted, system-wide effort, such as the International Geophysical Year was to the earth and related sciences. "All research efforts in Antarctica," says oceanographer Sayed Z. El-Sayed of Texas A&M University, "must be done within the ecosystem concept."

If the health of the ecosystem is to be maintained in the face of imminent exploitation, says El-Sayed, overall biological productivity must be preserved. But to do this, the most fundamental knowledge gaps—involving the population dynamics, physiology and behavioral patterns of the major species in the system—must be filled.

Despite such sentiments, widely held by American and other antarctic biologists, an estimated 50,000 tons of krill are already taken annually and marketed as food products, chiefly by the Soviet Union, Poland, Japan and, most recently, both East and West Germany. Additionally, Chile, which earlier tested the krill catch and processing technology, is expected to return, and several nations are reportedly considering the construction of on-site processing plants.

Uncertainty about the antarctic marine ecosystem, however, could modify some of this activity—or at least its success. British scientists had long reported that there are good and bad years for krill productivity. Both British and West German scientists have reported to their American colleagues that this past season was a bad one in the traditionally krill-rich waters around South Georgia and the South Orkney Islands; Soviet, Polish and East German trawlers, seeking fish or, in their absence, krill, were all reported to have been disappointed and to have left the Scotia Arc fishery with considerably less than full holds.

The British apparently don't believe that past heavy harvesting seasons would have depleted the stocks; cold austral winters are assumed to presage reduced populations for the subsequent summer, and the 1977 winter was a particularly cold one.

Mary Alice McWhinnie of De Paul University in Chicago concedes the correlation, but believes not enough is known of the antarctic euphausiids to reach any conclusion. She believes there may be races of them that vary among each other in such characteristics as longevity. Euphausiids that live two years, for instance, may be large enough to harvest after a year. But those that live four years would take longer to mature; if the heavy harvests of recent years were of such a breed, this year's stocks would not have grown enough to make harvesting feasible, she suggests.

A unique ecosystem

Most of life in the antarctic waters is unique to the area. It flourishes south of the Antarctic Convergence, but only a few antarctic species are found elsewhere in the world. This is attributed in part to the geographic and climatic isolation of the continent and its surrounding seas— an isolation that has been in effect for tens of millions of years.

As the South Pole area became colder over time (see "The Earth Beneath the Poles," in this *Mosaic*), organisms that survived adapted exceptionally well, with the effect that small numbers of species tend to dominate sharply. There are, for example, five species of the genus *Euphausia* in Antarctica: *E. valentini*, *E. crystallorophias*, *E. frigida*, *E. triacantha* and *E. superba*. But it is the *E. superba* that dominates. No one has ever been able to put a number to this dominance, says McWhinnie, but the other species, and genera other than *Euphausia*, are represented by substantially smaller populations.

Fish species, too, show a high degree of endemism, and three-quarters of the antarctic benthic fishes belong to one group: the nototheniiformes. As abundant as they are, these fish are not generally regarded as an exploitable food resource; they are too small and bony. Fishing for the more familiar cod and halibut takes place in the sub-antarctic waters of the South Atlantic and South Indian Oceans, but, in the Antarctic itself, there are few familiar or usable food fish. About 80 percent of the fish found in the Antarctic have not been encountered elsewhere.

The most abundant seal in the world, the crabeater, accounts for 85 percent of antarctic pinnipeds; more than 99 percent of the bird biomass are penguins, and the penguin population itself is dominated by the Adélies. Until recently, the largest whale population in the world, the baleens, vastly overshadowed the rest of the antarctic whale population. Though they still dominate the toothed (principally sperm) species, the numbers are not nearly so impressive.

This high degree of endemism makes the antarctic ecosystem more vulnerable than other ecosystems. "The organisms are limiting their own ability to survive by being bound to an area," says El-Sayed. "Once an organism is not bound to an area, it can move about for eating and reproduction. But if it is geographically limited, and then something happens to the environment on which it depends, the effects could be disastrous."

El-Sayed is concerned because it is not clear, given this limitation, whether or not antarctic species could survive abuse: "oil exploration and mining efforts in the Antarctic, for example. An oil spill or hydrocarbon leakage would disrupt many parts of the ecosystem. The surviving system would be totally changed. The system is a simple one," says El-Sayed,

Antarctic marine life. Food-chain relationships among inhabitants of the pack-ice zone of the antarctic marine ecosystem (top). Leopard seals have recently come to be known to include crabeater seals in their diet. The living resources of the southern oceans (below) produce a relatively short food chain.

S. Z. El-Sayed

Short food chain

The high productivity in the Antarctic—as in other rich marine areas of the global oceans—is directly related to phytoplankton production. "In a few restricted areas of the world," says John H. Ryther of the Woods Hole Oceanographic Institution, "surface waters are diverted offshore and are replaced by nutrient-rich, deeper water. Such areas of coastal upwelling are biologically the richest parts of the ocean."

Phytoplankters depend on sunlight and the nitrogen and phosphates brought up from the bottom by the upwelling (see "The Sea Turns Over," *Mosaic*, Volume 5, Number 1). Although phytoplankton has a severely limited growing season in the Antarctic because photosynthesis can take place effectively during only half the year, it accounts for 80 to 90 percent of the primary production in the marine ecosystem; it also produces food chains that tend to be very short—and efficient: Krill feed directly on phytoplankton; crabeater seals, Adélie penguins and baleen whales feed directly on krill. "There is a higher yield of animals when they feed directly on the phytoplankton," explains Ryther (see "Probing the Bering Sea Shelf," in this *Mosaic*). "In the middle of the ocean, a species might have to go to six links removed from the primary source of production for its food, losing nine-tenths of the nutritional value with each link."•

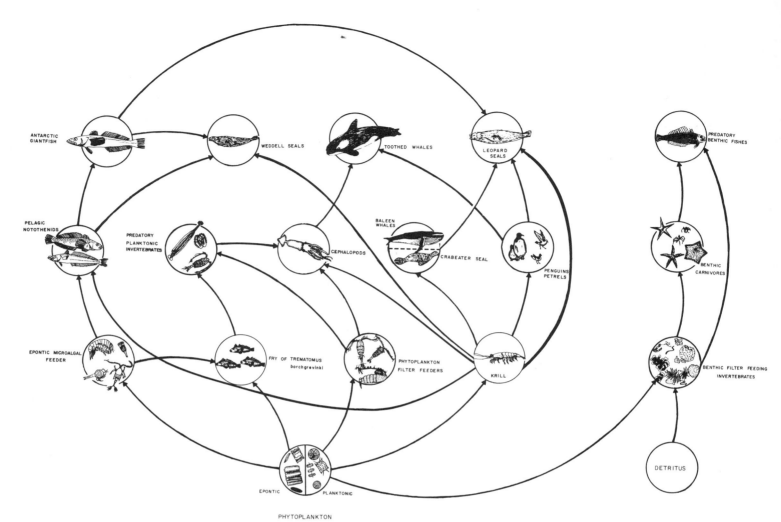
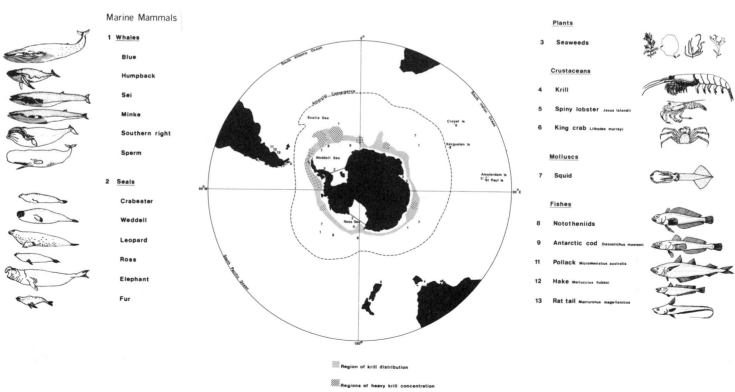

"weakened by its dependency on one organism for millions of years."

Krill genotypes

"There is little that people agree on about krill—except its taxonomic name," says Charlene Denys, a doctoral candidate at De Paul University who is working with McWhinnie on Antarctica's most prolific zooplankter.

Studies made in the late nineteen-twenties and early thirties, Denys and McWhinnie recount, reported krill to be everywhere in the rich antarctic waters. But more recently, researchers have found that, in fact, the southern oceans are not uniformly teeming with krill. There are high and low spots of productivity.

Such possibilities as geographic separation and the possible longevity differentiation have led McWhinnie's group to consider whether or not there are different genetic types or races within the species *E. superba*. "If our laboratory methods are correct," says McWhinnie, "the disparities in the literature about longevity could be interpreted to be the result of different genetic stocks. But they could also be due to different criteria of age, different food levels, temperature or other environmental factors."

If the differences identify different euphausiid stocks, McWhinnie says, "They might express their differences by sexually maturing at different ages, and by each having a different enzyme complex." These, she adds, are only assumptions, but their verification could have a sharp impact on what stress may and may not be imposed on the system with impunity.

Back in their De Paul laboratory after spending the last austral summer in Antarctica collecting krill samples, the McWhinnie team embarked on genetic studies to determine if there are indeed different races of krill and, if there are, to learn more about the differences in the fecundity, spawning, growth rates and longevity of the genotypes.

Their first major study involves obtaining a biochemical fingerprint of euphausiids to see what the similarities and differences are. Early results of enzyme analysis indicate the presence of at least two different genetic types from the area, on the west side of the Antarctic Peninsula, that the De Paul team normally visits.

Comparing overall growth rates in samples from single sites is another way to test for genotypes: "There has been a tendency to average growth statistics on the assumption that krill make up one population," says Charlene Denys. "But now there is an indication that some krill are considerably larger than others at an earlier stage of development, and even that some krill can breed more than once. In fact, some might even survive through several annual breeding cycles." The inference from these genetic studies seems clear: If there are different genotypes of krill throughout the antarctic seas, harvesting regulations would have to be tailored to best fit and ensure the survival of each of them. Premature exploitation of multiple breeders, for instance, could have wide-ranging effects on the productivity not only of krill but also of the entire ecosystem.

During the 1977-1978 southern summer, more by luck than by executed design, some of the krill the De Paul team brought back for sampling included females carrying not only eggs, but also sacs of sperm—two each—placed by a male on the female's thelycum, a genital pore or pouch. (Krill were also found—both male and female—"with sperm sacs stuck all over them." Ordinarily, the male

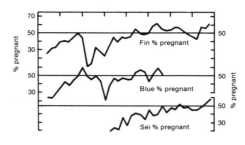
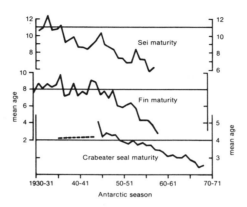

Secondary changes. Changes in the maturation and pregnancy rates of mammals in the southern oceans shifted markedly when the hunting-reduced population of baleen whales left "surplus" krill available to other krill eaters.

R. M. Laws, *Philosophical Transactions*

of such creatures carefully deposits sperm on the female's thelycum; euphausiids, with overlapping hormonal attributes between male and female to make distinctions, even by other euphausiids, difficult, apparently go on a wild fertilization spree—the males hitting everything within range—to ensure proper deposition of enough sperm to foster the species' perpetuation.)

Those captured females with properly pouched sperm, back in the De Paul laboratory, were placed in jars holding seawater kept at a comfortable and familiar one- to two-degree-centigrade temperature. When each female released her eggs they became fertilized (at an average of 2,000 to 3,000 per spawn) and, within three to four days, the eggs hatched.

So far, little is known about the development of krill, except that they go through 12 larval stages from the egg to the juvenile. Although McWhinnie's young krill died of a bacterial infection after surviving only three larval stages—18 days of life—the researchers made round-the-clock observations and documented the growth and development of half-millimeter krill eggs into millimeter-long krill larvae. (At the end of a year and its 12 larval stages, a juvenile krill is usually 11 to 13 millimeters long; an average adult krill can measure 40 millimeters.)

Aggregation patterns

Part of what has made the antarctic krill so popular a fishing target—especially since so many nations have reserved to themselves fishing rights out to 200 miles from their coasts—are what appeared to be the krill's especially accommodating "swarming" patterns (see "All That Unplowed Sea," *Mosaic*, Volume 6, Number 3). They seemed to mass on the surface at night for easy scooping up by trawlers. (The best krill fishing was, until last season, the Scotia Arc area around South Georgia and the South Orkney Islands.) Swarms have been known to be vast, as much as three meters deep and several hundred meters across, though most are considerably smaller.

But so attractive to trawlermen was this apparently accommodating behavior, and so characteristic of krill did it seem to be, that a catalogue of observations, and speculation about an apparently uniform set of patterns, came to be generally accepted:

• When krill are at the surface, actively feeding, apparently they disperse. They join together as a swarm when they stop feeding and are ready to descend. The onset and duration of a swarm formation is directly related to food.

- Large swarms make krill an easy catch (but it is not known what proportion of the total krill population is present in the swarm or what is the absolute density of the swarm).

- There are krill that swarm and krill that don't. Exploiting the swarmers won't threaten the species; loners will fill the gap. (This assumes that all euphausiids are like all other euphausiids; if you've seen one, you've seen them all.)

- Krill apparently respond to light: The densest swarms are found at the surface on the darkest of nights; as daylight's intensity increases, the animals—together—migrate downward, returning to the surface in 24-hour cycles. (The Research Vessel *Hero* operates at night now, McWhinnie says, so the scientists don't have to fish so deep for samples.)

McWhinnie, in her intense search for clues to krill life cycles and habits, is seeking especially for answers to the question: Do different patterns indicate different krill genotypes, perhaps of different longevity?

She returned from her last voyage, to the Bellinghausen Sea with new insights, credited in part to other scientists, including Ole Mathiesen of the University of Washington in Seattle and David Cram of the Union of South Africa, to the effect that "swarming" is not a single phenomenon. In fact, there seem to be three patterns, though not much more is known of the forces that drive them or the population differences, if any, among them. Nevertheless, McWhinnie declared in a recent letter to El-Sayed, "The word 'swarm' will now have to be qualified by those who use it to specify which of the three types of aggregations that are encountered is intended.... These I will describe in the literature and will hold them as caused by different factors until more study requires further refinement and corresponding modifications in swarm classifications...."

As she describes the swarm classifications, they are:

1) A relatively rare, aligned and tightly packed arrangement of euphausiids, all moving, in concert, as a single body. Probably all adults, possibly of a single sex. Extent and depth are unknown, but they are "what anyone envisages when the word 'swarm' is used.... These would not be found on a sonar echogram except if experimental studies arranged the transducer above the water level."

2) A random arrangement of members, slightly separated from each other but in relatively high density over a large area, "...as far as the eye can see...." These can be seen in krill-rich areas at any time at night. They are not really swarms, but are a "layer of high numbers" brought to the surface with declining light at dusk or driven to as much as 100 meters depth with daylight.

3) Aggregates of krill seen often on conventional echograms of "spots." These are probably the ones most commonly fished commercially or sampled experimentally "by all who search for krill via sonar."

Too many krill?

"The reason for all this research work in krill," says McWhinnie, a 16-year veteran of Antarctic research, "is that we have a chance to understand a vital organism in an ecosystem before it is harvested."

Consider the issue of the krill "surplus," she says. It has been estimated that, since the baleen whale stocks were reduced by some two-thirds, and barring some food or other limitation to krill proliferation, there are tens of millions of tons of krill each year that are no longer eaten by the whales—extra krill that could easily be harvested without disturbing the ecosystem, or even that *must* be harvested to keep the system in balance.

Marine biologists like McWhinnie, however, argue that there is no krill surplus. Instead, an adjustment had been and is taking place within the ecosystem, and the "extra" krill are being preyed upon by many predators rather than one.

There is evidence, for instance, that crabeater seals—big krill consumers—are coming into maturity more than a year earlier than usual, and that elephant seals—which do not eat krill but which eat squid, which eat krill—are also growing at a faster rate. Surviving baleen whales are also coming into maturity at an earlier age, and there have been significant effects on one skua population.

Each of these is explained by the fact that reproductive success in a marine system is sensitive to nutrition; the more krill there are for seals, whales, squid, skua and penguins, the better their reproductive success will be and/or the faster they will reach reproductive maturity. For the baleen whales, however, the explanation might also have to encompass the fact that, in a depleted population, reduced competition from one's own kind—or just more space in a pod—might be factors too. And, as for overpopulation of the beneficiary species, some management-oriented fisheries biologists are suggesting that the penguin populations might have to be cropped to keep them from breeding beyond the ability of a balanced system to support them.

Crabeater seals

Although krill are a central factor in maintaining the health of the antarctic marine ecosystem, studies of other inhabitants—specifically marine mammals and birds—are important in determining the biological fate of the area. Armed with knowledge about the broader functioning of the ecosystem, scientists can then begin to predict possible changes that could result from the ultimately inevitable harvesting of the marine resources.

The vast biomass of crabeaters, for example, is of potential commercial value for leather, oil and animal food. The seals are difficult to harvest, however, because, being a pack-ice species, they are not easily accessible. Not only do they rarely come close to shore, but they also tend to stay in small groups and are not found in the more harvestable clumps that make their arctic cousins so vulnerable. A detailed understanding of population regulation mechanisms could help ease the transition between their present, unexploited condition and controlled exploitation, according to biologist Donald Siniff of the department of ecology and behavioral biology at the University of Minnesota.

As an example of this, Siniff observes, when crabeater pups develop into yearlings, they come in from their traditional pack-ice homes to live near shore during the breeding season. The adults remain on the pack ice and exhibit very aggressive mating behavior. "The discovery of large numbers of young seals near shore in the Antarctic Peninsula," Siniff explains, "could give rise to guidelines...to prohibit and prevent exploitation of the young in certain areas and at certain times of the year. If seals are going to be harvested again, the consequences must be known."

Siniff and his colleagues have been spending austral summers at Antarctica's McMurdo Station for a decade, observing the seal populations. One of their projects involved tagging female Weddell seals and calculating how many young they have. "Female seals," he explains, "can have young every year, but about half of them don't. We are trying to figure out why the population goes up and down."

Siniff's group is hypothesizing an

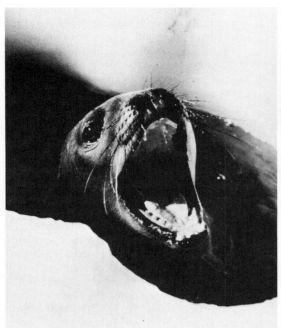

Antarcticans all. A Weddell seal (top left), an Adélie penguin, a crabeater pup and mother and a south polar skua are all part of the biomass, the integrated study of which is strongly advocated.

D. R. Cline; U.S. Navy; J. L. Bengtson; David Parmelee

interaction between the population and the environment. Weddell seals, for example, have their pups near fissures in fast—anchored—sea ice, where the female has easy access to the water. As these regions become crowded, fewer pups are produced; conversely, more pups are whelped when population densities are substantially reduced.

Predators, too, can affect population levels, but, until recently, the seals' natural enemies had not been properly identified. Eighty percent of observed adult crabeater seals, for example, have massive scars, which are presumed to be the result of attacks by predators. Since there are so many badly wounded crabeaters, it is likely that many of the seals did not survive the predators' attacks; Siniff has found that the male protects its pups only for a short time—while it hangs around the pack ice waiting for the pups to be weaned and for estrus to begin—making it unlikely that the scars are earned in defense of young or mate.

Before this pattern was ascertained, the crabeaters' scars were thought to be caused by killer whales, and there were numerous reports to this effect in the scientific literature. But recent observations by Siniff and his colleagues place the blame on the leopard seal—the biggest and most ferocious seal in the Antarctic. Although leopard seals were long known to eat penguins, their voracity was not fully appreciated; they apparently will take on almost anything reasonable. "We didn't realize the diversity of food available to the leopard seal," says Siniff, "but it is important for us to know this, to get a better understanding of what is going on in the ecosystem."

Skuas, terns and gulls

Skuas and gulls are also prolific in the Antarctic Peninsula area; they have one of the highest rates of productivity of any skua or gull in the world, according to David Parmelee, chairman of the field biology program at the University of Minnesota. The high bird productivity is thought to be directly associated with the high productivity in the seas.

Parmelee's group of researchers has been following the eating habits of two species of skua (a web-footed, rapacious, high-latitude member of the jaeger family and close relative of gulls). Parmelee has followed the brown skua, which lives mostly on penguins and finds its food near shore in the area of rookeries, and the south polar skua, which flies out to sea to forage for krill and fish. At Antarctica's

> **Through the Ross Ice Shelf**
>
> Researchers working in the austral summer of 1977-1978 melted a cylindrical hole through more than 400 meters of Antarctica's Ross Ice Shelf with a supersonic flame jet drill (see "Tales the Ice Can Tell," in this *Mosaic*). They opened an undersea world never seen before, as cameras and scientific equipment were lowered through the hole to study both the ice shelf and the 200 meters of water beneath it.
>
> Some scientists had speculated that there would be no life there because there was no source of light for photosynthesis. Others felt that life under the shelf would be a similar— but sparse—replica of the antarctic marine ecosystem, based on the expectation that currents would carry in secondary food sources.
>
> And, indeed, that is just what researchers found: fish, shrimp, euphausiids, amphipods and isopods normally found in the waters of the southern oceans, but in much smaller numbers. Many amphipods were noted to be brooding young and carrying eggs, suggesting an environment, food included, to support some organisms through entire life cycles.
>
> Life under the Ross Ice Shelf, however, is different from life outside it. Under the shelf, despite a variety of species, population sizes appear to be very low. As a result of this, the research findings are more relevant to life in the benthic depths of the ocean (see "The Deep Seas—Unexpectedly, an Astounding Variety of Life," *Mosaic*, Volume 7, Number 3), explains Jere Lipps of the University of California at Davis, who coordinated biological studies at the shelf site. There, too, many species are represented in an environment apparently unable to support many members of any one species. It is another example, however, of the ability of life to adapt to what might seem to be the harshest, most forbidding conditions. •

Palmer Station, they recently recorded an explosion in productivity in the south polar skua populations, presumably the result of the increased availability of krill in the seas, while the non-krill-eating brown skua remained stable.

Recently, the area around Palmer Station remained icebound most of the summer, Parmelee reports, and the researchers had a chance to observe what happens when the south polar skua is cut off from the krill. The results were dramatic: The skua did not breed effectively; in one year, Parmelee says unequivocally, its productivity fell from a record high to zero. The brown skua, in contrast, fed well and bred as usual.

Other birds were also affected by the icebound conditions. The antarctic tern, with surface eating habits similar to those of the south polar skua, neither foraged normally nor bred successfully. And the southern black-backed gull, which feeds mainly on limpet at the edge of the sea, with its source of food mostly under the ice, also showed near-zero productivity. In contrast, other bird species, such as penguins and shags, which traditionally dive deep for food sources, even those temporarily locked under the ice, were able to feed and breed normally.

Exploitation, uncontrolled...

These studies and hundreds more on marine species are being pressed—in advance of man's further intervention in the ecosystem, and in clear anticipation of it. Biologists feel with increasing urgency that efforts must be made to know how the system works and how it will respond to pollution and to exploitation of its key organisms.

Previously, exploitation took place without concern for the ecosystem. Fur hunting began, for instance, soon after British explorer James Cook reported in 1775 that the South Georgia area was a rich source of the antarctic fur seal (*Arctocephalus gazella*). Uncontrolled sealing continued throughout the 19th century, leading eventually to the almost total extermination of the species. Sealers were forced to look for other resources—and they settled on the elephant seal, *Mirounga leonina*, for its oil value rather than its hide. Because profits were considerably less, the industry became uneconomical before exploitation had reduced the elephant seal stocks to as critically low a level as the fur seal. (The fur seal population on Bird Island, South Georgia, has been coming back dramatically. It has grown from 100, including pups, in 1936, to 350,000 in the last census, in 1977, with annual pup production now thought to be about 90,000.)

Learning little from the impact of seal exploitation, whalers came into the Antarctic in the early 19th century with their harpoons and small boats, towing their catch to shore stations for rendering. A century later, whaling factory ships roamed the seas, processing and storing all that is useful from a 500-kilogram whale in less than an hour. These boats were so efficient that the whalers could hunt indefinitely without a need to return to land, and could thus range over all the southern oceans in search of their prey. The results were predictable: Of the baleen whale population, the blue, fin, humpback and southern right whales declined drastically; the smaller sei and minke whales are less profitable but are the hunter's present target. A toothed whale species—the sperm—is also found in the Antarctic but has always been outnumbered by baleens.

...and regulated

Whaling is now regulated by the International Whaling Convention, and limits are set on total catches. Harvesting of seals is also being regulated. Krill—the latest in what could be a long list of exploitable marine resources from the Antarctic—may now be coming under international controls. Thirteen nations have proposed an Antarctic convention that would control, through a commission on the conservation of living marine resources, the anticipated mass harvesting of krill. The convention would also set the stage for tackling the other controversial issues surrounding exploitation of Antarctica's resources— including suspected deposits of oil and coal (see "The Earth Beneath the Poles," in this *Mosaic*).

Meanwhile, scientists from more than a dozen countries have joined forces for a Biological Investigations of Marine Antarctic Systems and Stocks (BIOMASS) program. Their goal: to gain a deeper understanding of the structure and dynamic functioning of Antarctica as a basis for further management of living resources.

Thus, BIOMASS may be the latest, but certainly not the last, large effort to understand the fragile antarctic ecology.

Just as Rachel Carson's *Silent Spring* said things that scientists knew, but said them in a voice ringing enough to arouse popular concern (see "Crops Are Ecosystems Too," *Mosaic*, Volume 9, Number 4), scientists like McWhinnie, El-Sayed, Parmelee, Siniff and others await a clarion voice to call attention to the ecological problem of their concern. There is, apparently, still time to learn the antarctic marine ecosystem so that the inevitable exploitation of its resources can be managed properly. But, the scientists warn, time, for Antarctica, may rapidly be running out.•

The National Science Foundation's support of the research reported in this article has been principally through its Polar Biology and Medicine program.

Probing the Bering Sea Shelf

An integrated study of an arctic marine ecosystem finds that not all theories hold in such an environment.

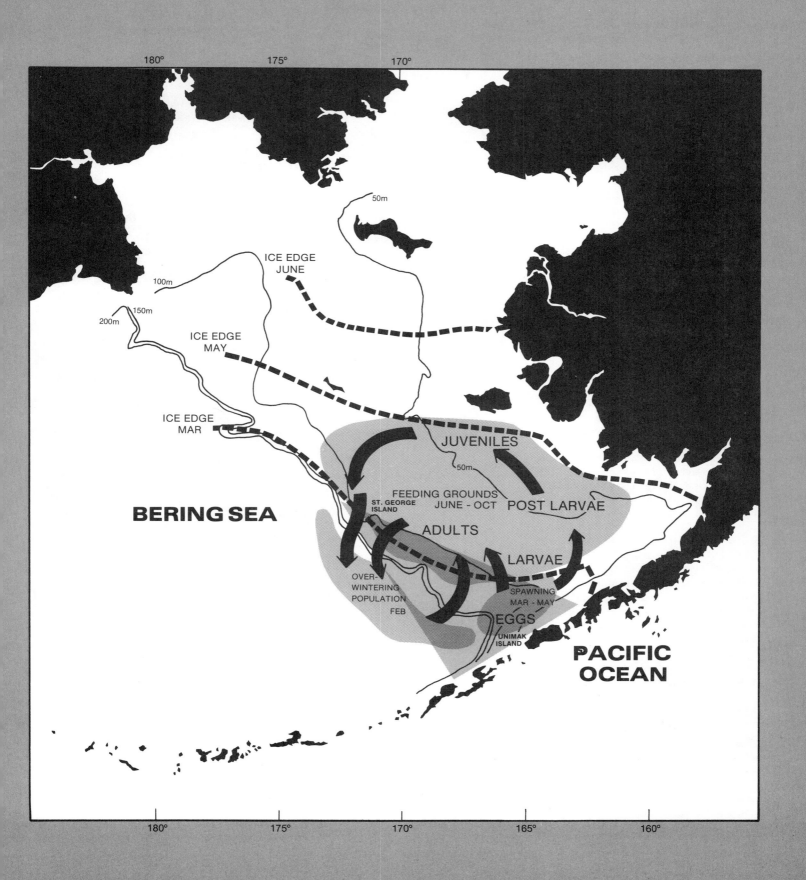

The shelf under the Bering Sea once formed the land bridge linking the Americas to Asia (see "The Earliest Known Americans," *Mosaic*, Volume 6, Number 2). And for five months of every year, when the shallow sea out to the edge of the shelf freezes over, there is still such a bridge, complete with continental climate enough to provide habitats for such land animals as foxes and owls.

During the rest of the year, however, the Bering Sea shelf, extending from the Aleutian Islands past the Pribilof Islands and northwest toward the Soviet Union, and overlying about half of the three-million-square-kilometer continental shelf of the entire United States, harbors one of the world's richest fisheries. The fishery, supported by a productive and extraordinarily efficient marine ecosystem, amounts annually to 2.5 percent of all the world's marine animal harvest. And this accounts only for the harvest taken by humans. The fishery is actually divided almost equally between human consumers and a large population of indigenous consumers, including whales, walruses, fur seals, polar bears and arctic birds. These creatures—like humans, top predators or feeders at a trophic level near the top of the food chain—crop about two million tons of Bering Sea fish each year. In addition, the fishing industries of Japan, the Soviet Union, Korea and the United States have, in a single year there, gathered as much as 2.3 million tons of seafood for human consumption.

The king crab and snow crab are an increasingly valuable segment of the fishery on the American market. But the largest single source of commercial Bering Sea protein, which for the most part finds its way into the dinner pots of Asia, is the Alaska pollack, a half-meter-long member of the cod family.

A theoretical problem

The Alaska pollack, which constitutes three-fourths of the biomass gathered in the fishery, is the most abundant upper trophic-level fish in the Bering Sea. As such, it commands the attention not only of the fishing industry but of marine ecologists as well: Investigators studying the marine ecosystem of the Bering Sea

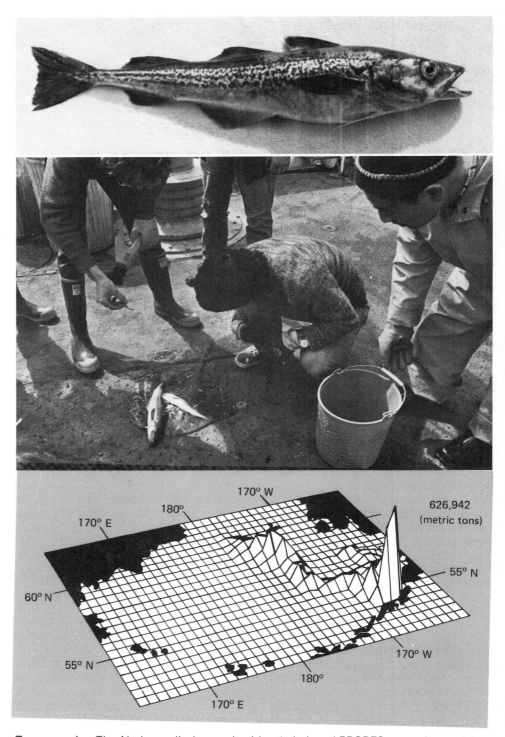

Tracer species. The Alaskan pollack, examined (center) aboard PROBES research vessel, is a major resource of the Bering Sea shelf. Earlier catch data (bottom) and an intense research effort contributed to profile (left) of pollack's life cycle.

National Oceanic and Atmospheric Administration; Barry McWayne

shelf in a major current effort are focusing on the Alaska pollack as a tracer species, the growth and development of which characterizes for them certain critical processes of the Bering Sea shelf ecosystem. Their study is called PROBES, an acronym for Processes and Resources of the Bering Sea Shelf. (See "PROBES evolves," accompanying this article.)

While the economic importance of the pollack would be more than enough to justify the research, the actual role of the pollack in the study is more as tool than target; the theoretical problem into which the pollack figures is that the Bering Sea produces more upper trophic-level biomass than currently accepted ideas of trophic dynamics encompass.

The accepted notions, propounded in 1942 by the late Raymond L. Lindeman, hold that only about ten percent of material and energy in a food chain can be transferred upward at each step in the chain. Thus, in any marine ecosystem, primary producers, algae and other photosynthetic phytoplankters, should support a grazing population weighing in at only ten percent of the phytoplankton biomass. In turn, the next level of consumers ought to incorporate ten percent of the grazers' mass and energy, or only one percent of that fixed originally by the plants.

But the Bering Sea shelf (not unlike the waters in the Antarctic; see "Antarctica: No Catch Limit Yet," in this *Mosaic*) does not fit the model. According to C. P. McRoy of the University of Alaska, PROBES's chief scientist, jumps at each level of what must be a short, tight food chain in the Bering Sea appear to be much greater.

"The Bering presents an interesting paradox," says McRoy. "The region is renowned for its populations of birds, seals and fishes—all higher trophic-level consumers. On the other hand, our knowledge of the productivity of the phytoplankton indicates that the Bering Sea is not particularly impressive at the primary trophic level. The answer must be in an efficiency of energy and matter transfer that is considerably above the ten-percent level."

These estimates, however, are

Biosamples. Samples dipped from the Bering Sea shelf area are concentrated and examined for numbers and kinds of organisms in laboratories aboard the RV *Thompson*, shown anchored at St. Paul Island in the Pribilofs in the Bering Sea, as a flock of murres, the northern equivalent of penguins, flies past.

Barry McWayne

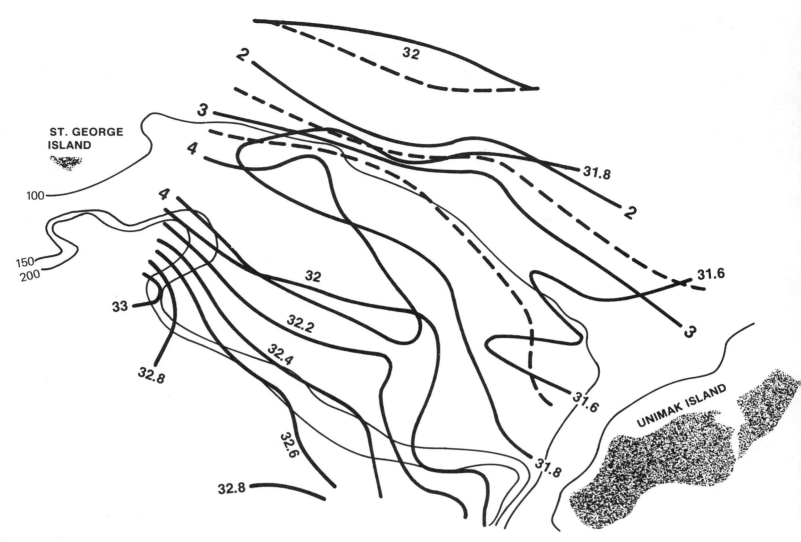

Physical oceanography. Bottom temperature (black) and surface salinity (color) profiles identify the major Bering Sea shelf fishery between Unimak and St. George Islands (above). Temperature vs. depth profiles (right) reveal the fine-structure mixing or interleaving of oceanic and shelf waters.

L. K. Coachman

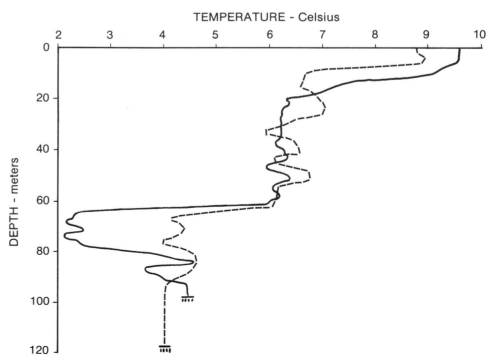

preliminary; the study, though it has already made significant contributions, completed in the spring and summer of 1978 only its second full season. National Science Foundation support for the project, scheduled to last for six seasons, began in 1976. In 1981 the study will have reached a point at which it will be possible to synthesize the essential features of the Bering Sea shelf ecosystem in a computer simulation model.

The pollack hypothesis

The theoretical base of all the PROBES investigations relates to a focal hypothesis concerning the survival of the pollack. The hypothesis is that the survival of a given year's group of high trophic-level consumers, including the pollack, depends on the vast size of the moderate-depth shelf and the sequence of certain physical and biological events on that shelf. "A specific combination of events *must* be responsible for the high regional productivity," says McRoy. But the collapse of the hypothesis would not wipe out the project. "Any test sufficient to disprove our basic hypothesis would provide sufficient input for new hypotheses based on sounder evidence. When you work with a complex dynamic system," McRoy goes on, "you have to retain the flexibility to accommodate changing ideas during the course of the work."

Within the main hypothesis is an array of working hypotheses or assumptions to be tested. One major assumption was shown to be incorrect by the end of the first two seasons of data gathering: "What had been thought about the Bering Sea," says McRoy, "was that it harbored a counterclockwise current, moving in a pattern northeastward along the north shore of the Aleutians, then northwestward to the area of the Pribilof Islands and finally southward again toward the Aleutians.

"In this circuit, we thought, the life history of the pollack was played out. We originally envisioned the spawning of pollack in the south along the Aleutians, the maturing of the pollack during a subsequent season along the northwestward flowing part of the current, followed by the return southward of mature pollack which would spawn in the southern Bering Sea to begin the cycle anew."

A negative finding

Much of the excitement of PROBES's first two seasons, however, lay in the discovery that there is no circular current involved in the life history of the pollack. In fact, there is no important current at all in the productive region of the Bering Sea. This important realization—negative though it was—opened the way to alternate sources of water movement with implications far beyond those of the

---

PROBES evolves

The roots of the study of the Processes and Resources of the Bering Sea Shelf (PROBES) date back to the early nineteen-seventies. The idea was conceived at a multinational symposium on Bering Sea oceanography held at Hokkaido University, in Japan, in 1972. Preparation for the experiments began the next year at a workshop instigated by D. W. Hood, then director of the University of Alaska's Institute of Marine Science and now principal investigator of PROBES. Besides NSF, agencies involved in the workshop included the Alaska Sea Grant program of the National Oceanic and Atmospheric Administration, the Food and Agricultural Organization of the United Nations, the U.S. National Marine Fisheries Service and the Alaska Department of Fish and Game.

By 1976 PROBES had evolved into a five-part program to explore the efficiency of the Bering Sea shelf ecosystem, using the pollack as a tracer species. The five phases of the undertaking are:
• Water circulation and mixing (important to the supply of nutrients to the ecosystem);
• Nutrient dynamics (the concentrations of chemical raw materials for algae growth);
• Primary productivity and phytoplankton (the lowest level of biomass formation);
• Upper trophic-level ecology (the pollack and the tiny animals on which pollack feed), and
• Ecosystem analysis and synthesis (with an eye toward predicting the consequences of disturbances in the ecosystem).

Centered at the University of Alaska's Institute of Marine Science, it involves scientists there as well as from the University of Florida, Washington State University, Brookhaven National Laboratory and San Francisco State University.

Data collection is a major task. The oceans in general are a research frontier, and the high-latitude oceans are especially so. "If PROBES alone were responsible for gathering all the data on which it depends," says Hood, "its work would not be as far along as it is today. Fortunately, there are other agencies with an interest in the Bering Sea shelf, and their data, which helped make it possible to plan the present ecosystem study, will continue to shore up certain aspects of PROBES."

The U.S. National Marine Fisheries Service, for example, which has a responsibility to manage the fishery of the Bering Sea, has pursued research toward a model of the higher trophic levels in the Bering Sea. Data have come from historical records of fish catches and from annual trawl surveys of the large animals in the ecosystem. But the Fisheries Service's concern, which is to analyze factors that promote a stable fishery, depends on the ecosystem's productivity; PROBES should benefit that effort.

Another source of data for PROBES is the Outer Continental Shelf Environment Assessment program under which the Department of Commerce's National Oceanic and Atmospheric Administration has sought to describe the physical, chemical, biological and geological aspects of the shelf environment for the Department of the Interior's Bureau of Land Management. (A principal motive for that effort is to learn what portions of the shelf ecosystem may be vulnerable to oil pollution and side effects of oil exploration.)

Research vessels

PROBES's own floating data collectors include the research vessels *Acona*, operated by the University of Alaska, and *Thompson*, operated by the University of Washington. The *Thompson*, 68 meters long, is capable of withstanding storms that are characteristic of the spring and fall in the Bering Sea. The smaller *Acona*, the only ship available to PROBES during 1976 and 1977, must stop working and take cover during severe weather. But mixing of waters during a storm may be an important component of the ecosystem, and the *Acona*'s limitations were a crucial defect in the early data gathered.

But even the *Thompson* cannot sail into ice, and the ice cover of the Bering Sea is one of its interesting features. Extending each winter out from shore just to the edge of the continental shelf, it covers the deeper, more productive part of the ecosystem that is of greatest scientific interest. The underside of the ice is a veritable incubator for algae and may provide a refuge for young pollack as well. When it melts, it releases low-density water into the surface layer, along with large quantities of algae that have grown with the ice.

Because a close look at the ice-covered ecosystem could reveal some interesting facts, Hood hopes that a proposed reinforced research vessel will soon be available. Not an icebreaker, the ship would nevertheless be able to sail into the ice-covered shelf. Sea ice does not form continuous sheets but is usually broken by leads; an ice-capable vessel would be able to steam a useful distance into the ice. •

Bering Sea study alone.

There are many possible mechanisms by which nutrients and planktonic creatures may be transported in an ocean system, explains L. K. Coachman of the University of Washington, a physical oceanographer. It was Coachman who made the initial determination of the absence of currents and who led the search for alternate sources of water mixing and nutrient transfer in the Bering Sea system.

Advective forces for the movement of water, such as currents, tides and winds, are the usual basic nutrient transfer mechanisms; without them, the surface layer of a body of water may, unless some other mechanisms can apply, become depleted of its nutrients by standing phytoplankton masses. Masses of water in motion in relation to each other because of variable local currents might be such another mechanism. It could cause waters to mix as a consequence of turbulence at the boundary between them. Eddies, for example, may form and spin water off both masses at the boundary layer; mixing as well as eddy diffusion can occur, moving nutrients and other matter from regions of greater to regions of lesser concentration. Such nonadvective motions, then, might also bring chemical nutrients near the surface, where phytoplankton can return them to the nutrient cycle.

Still another source of relative motion can be density differences; a layer of water may be colder or more saline—hence denser—than the layer beneath it. As it sinks, its motion can give rise to mixing of the two water masses.

"But in the Bering Sea," explains Coachman, "density differences do not seem to be important. Waters that come from the deep ocean are more saline, but also happen to be warmer; these waters have about the same density as the colder but less saline inshore waters."

Since the Bering Sea lacks both major currents and density differences that might make mixing energy available, the answer had to be found among the other possible mechanisms.

Tides, winds and eddies

The mixing mechanism that Coachman has perceived on the Bering Sea shelf gets its energy not from a current, as originally envisioned, but from the tides and the winds.

In the Bering Sea, wind provides the energy for mixing water in the upper layers; tides provide energy to mix water in a layer near the bottom.

"The most important source of resupply of nutrients to the shelf ecosystem," says Coachman, "is the near-bottom water, coming in from the oceanic [offshelf] environment." The tide, he explains, functions as a kind of pump that draws this deep-sea water, relatively rich in nutrients, onto the edge of the shelf. The tide moves the entire water column, of course, but the lowest tier of water is slowed by friction with the bottom and made turbulent. Eddies then form and cause deep-sea water to mix with nutrient-poorer shelf water and remain behind in shelf water as the tide recedes.

With each tide, Coachman says, the cycle recurs; "By a kind of ratchet action, nutrients are replenished in the bottom layer of the shelf water where boundary effects provide energy for mixing."

Tidal mixing operates for about the bottom 50 meters of Bering Sea shelf water, says Coachman. The other kind of mixing, wind mixing, affects the upper 20 or so meters. The winds above the Bering Sea are variable, blowing sometimes out to sea, sometimes landward. This creates an opportunity for exchange of nutrients and other matter that is comparable to the exchange occurring deep in the water column with tidal movement.

If, for instance, the wind were driving ocean water across the continental shelf edge toward the land, a boundary layer would form between the moving surface layer and the water beneath. Where turbulence occurs, at the boundary layer, energy is made available to mix water and foster diffusion of nutrients into the shelf water. Later, the upper layer of water will again blow out to sea. The cycle is likely to be repeated several times each season, with new ocean water entering the shelf ecosystem to deliver a quantity of nutrients at each landward shift in the wind.

As Coachman and his colleagues describe the shallow inner part of the Bering Sea, where water depths seldom exceed 50 meters, the mixing actions of tide and wind overlap; wind mixing extends down to about 20 meters; tidal mixing influences the entire water column.

A sharp gradient

But this region is not the richest part of the Bering Sea shelf ecosystem, and Coachman and his colleagues have also come up with some details about water mixing in the deeper region, outward from 100 meters depth, where the fishery exists and where the marine ecosystem is most highly productive.

In water of depths of 50 to 100 meters, Coachman and his colleagues find, the wind and tidal mixing roughly complement each other, the one mixing from the bottom, the other mixing from the top. But along a line where the shelf water is 80 to 100 meters deep, a steep gradient occurs in the water's salinity. The salinity gradient marks a front or interface between the shelf water inshore and the regime of mixing and interaction in the region beyond the 100-meter isobath (a line along which all points are the same depth). The front was detected in 1977 by Coachman's measurements of the salinity of the total water column along a string of points, running from shallow water inshore out to the edge of the continental shelf.

Virtually all coastal waters are less saline than deep-ocean water; they carry fresh water that has run off the land. It should not be exceptional that a salinity gradient might occur over the Bering Sea shelf. But what *is* striking in the Bering Sea data, says Coachman, is the discontinuous character of the gradient: Salinity increases gradually out to about 100 meters; it increases more sharply from there.

The salinity difference is not great—from about 3.15 to about 3.23 percent salt content. But the rate of change between these concentrations is unexpectedly rapid. The gradient occurs over a distance of about 50 kilometers, from which point it levels off, holding steady for a distance of another 50 or more kilometers.

This is obviously a different regime from the waters closer inshore; the critical difference appears to be that the waters here are deeper; depths exceed 100 meters, Coachman explains, which means that the wind and tide are no longer able to stir the entire water column. From these intermediate depths in the region outward from the front, the data appear to bear out recent hypotheses about mixing of waters in such low-energy sea regimes. In the absence of effective tidal mixing or wind mixing, Coachman finds another mode of mixing on the so-called fine-structure scale.

A layered lake

When a temperature/salinity measuring probe is dropped into Bering Sea water beyond the 100-meter or so isobaths, the meter acts erratically. Readings seem to jump up and down as the instrument penetrates layers only a few meters thick, revealing what appears to be the layering or interleaving of waters of different origins. Coachman believes that the existence of the strata demonstrates an inherent tendency for two laterally juxtaposed waters of nearly the same density—shelf and oceanic—to interleave.

"The outer Bering Sea shelf appears now to be a natural laboratory," Coachman says, "in which to study fine-structure mixing. This mixing has been shown mathematically to be an effective way of transferring nutrients and material laterally through water in regions where kinetic energy for mixing, in the form of tides, current or wind, is not directly available."

The region of interleaving waters does not extend indefinitely out into the sea, says Coachman, though the interleaving layers do extend remarkably far, considering how thin they are. Only a few meters thick each, they may stretch on the order of 50 kilometers across the outer part of the Bering Sea shelf. Where the layering phenomenon ends coincides with the edge of the continental shelf, the shelf break. There the sea bottom drops away rather steeply to a depth of over 3,000 meters.

The shelf break also coincides with the outer margin of the fishery and the limit of the most productive region of the marine ecosystem. Moreover, along this same line is another salinity gradient of the kind found at the inshore front along the 100-meter isobath. And the shelf break or outer front shows a salinity gradient (the farther offshore, the more saline) of about the same steepness as the gradient at the inner front.

The picture the physical oceanographers are left with is of a kind of lake in the middle of the Bering Sea: an interfrontal zone between the 100-meter isobath and the shelf break. It turns out to be a physically defined zone that coincides with the rich ecosystem.

Thus, solid evidence now points to a large oceanographic regime that is clearly distinct from either shallow shelf or deep ocean and that is coextensive with the pollack habitat. It is an exciting picture to the investigators and accords well with the accumulating data from the other phases of their integrated study.

"The [physical] oceanographic data have refined our hypothesis tremendously," McRoy says. "We don't have fears any more that we'll miss any of the action by failing to locate pollack at a given stage of their lives. Now that we know they're distributed around a homogeneous habitat, we're sure to be able to observe enough to get a complete understanding of the system that supports them."

Nutrient dynamics

Closely linked to the oceanographic analysis of the Bering Sea shelf is the research aimed at nutrients: the chemicals, including nitrate, nitrite, phosphate, ammonium, silicic acid (a constituent of diatom shells) and carbon dioxide, that support plant growth.

Early efforts to gather data on nutrients were frustrating, report W. S. Reeburgh of the University of Alaska and T. E. Whitledge of Brookhaven National Laboratory. The samples they and their colleagues collected in the earliest stages did not store well and gave biased responses when analyzed ashore. Now the *Thompson* sails with an autoanalyzer aboard; chemical analysis can be carried out promptly as samples are brought up from depth.

The nutrient sampling so far has confirmed the physical description of the Bering Sea shelf given by the temperature/salinity data. "The so-called lake," Reeburgh says, "is like a chemical reactor, getting nutrients from all sides and driven by the energy of mixing. Nitrates are injected into the system from the seaward side; silicic acid and ammonium arise from the bottom where decomposition releases them."

The scientific group studying nutrients has found that the two frontal regions are very low in nitrogen-bearing nutrients and also low in carbon dioxide. The 1978 cruises have shown that nitrogen is low because of the intensity of growth of phytoplankton, which absorb nitrates as fast as they become available. As the season progresses, recycled nitrogen becomes available in the form of ammonium following the death, or consumption by grazers, of earlier generations of phytoplankton. This nitrogen supply, too, is rapidly exploited and held to a low level. By late in the summer, diatom growth has proceeded to a point where no further silicates are available either, and this phenomenon tends to bring the phytoplankton season to a halt.

As for carbon dioxide, Reeburgh and project director D. W. Hood plan to work on this question during the 1979 season by measuring the rate of exchange of gases between Bering Sea water and the atmosphere. To do this they will trace the concentrations of the inert gas, radon, which is produced by fission of radium atoms.

Hood and Reeburgh plan to observe the gradient of radon concentration near the sea surface. Radon, which occurs naturally, will tend to diffuse upward to the atmosphere at a rate governed by the thickness of the so-called laminar layer of surface water. The thin layer (on the order of 60 micrometers) has different physical characteristics from the rest of the water column, and it inhibits the transfer of gases between air and ocean. From the radon data, says Reeburgh, an inference will be drawn about the rate at which carbon dioxide passes into the system through the laminar layer. The rate of carbon dioxide exchange when related to the total amount of fixed carbon in the system will tell how much net increase is actually taking place in the ecosystem.

Primary producers

On the ladder of trophic dynamics, the rung above raw chemical nutrients is the growth of primary producers. These are the photosynthetic phytoplankters that use light energy to convert chemicals to living substance. Heading this phase of the study is John Goering of the University of Alaska, working with R. L. Iverson of Florida State and a group of graduate students. The phytoplankton study aims at quantifying the primary production on the Bering Sea shelf and apportioning the production among the several planktonic species that grow there.

One objective of the study of primary production, Goering explains, is to ascertain the nutrient or nutrients that limit the growth of phytoplankton: Which species is the first to be used up by the growing plant mass, or which one is supplied so slowly that it controls the rate of phytoplankton growth? The analysis of blooms figures into this question. A bloom is the rapid, massive growth of phytoplankton that may occur when good quantities of nutrients in the proper ratio are gathered at a spot under conditions suitable for phytoplankton growth. A bloom could feed a large population of grazing copepods and other animal (zoo) plankters, and these in turn can feed a population of recently hatched pollack. The couplings among such a series of events are at the heart of ecosystem analysis.

Especially the young

The final element in the food web being studied involves, of course, the pollack, especially the young, whose survival appears to depend upon critical sequences of biological events. The workers on this phase of the program include R. T. Cooney and Tsuneo Nishiyama of the University of Alaska and T. S. English of the University of Washington.

This group's work on upper trophic-level dynamics is in some ways the most

Zooplankton collectors. A pair of bongo nets lowered over the side of the RV *Thompson* brings zooplankton, eggs and larval fish aboard for laboratory examination.

Barry McWayne

difficult of the tasks. Pollack larvae are fragile and easily destroyed by sampling procedures that might otherwise be routine. Nevertheless, it is essential to find pollack eggs, locate the larvae and correlate the success of larval survival with other elements of the ecosystem: to determine what is present in the ecosystem that determines survival and growth rates of young pollack.

The life history of the pollack, says Cooney, is at this point only partially understood. Judging from acoustic or sonar soundings, it appears that adult pollack are found in patches throughout the interfrontal region of the Bering Sea shelf. "But this," Cooney cautions, "may not mean that the females spawn at random."

Evidence suggests that the outer part of the interfrontal zone is the scene of early spawning, in April, and that during subsequent weeks spawning fish are found progressively shoreward. Spawning continues until June. The eggs incubate for 10 to 30 days, depending on temperature, and hatch into 4.0-millimeter-long larvae.

When, after five to ten days, a larva exhausts the nutrients in its yolk sac, the young pollack, if it is to survive, must find itself in waters bearing the right sort of food: bite-sized morsels, including cells of large diatoms, copepod eggs and larvae. Though it is highly unlikely that any given pollack larva will survive, it is inevitable that somewhere in the interfrontal zone the timing of events will be such as to bring some pollack together with the right food.

According to Cooney, it will be essential to learn exactly what kinds of plant and animal plankters are present at particular times and depths in the water column, down to about 200 meters. "A hypothesis about pollack survival," says Cooney, "based on observations of good years in the anchovy fishery, suggests that storms that mix the water column to a depth of 20 meters or more may disrupt planktonic communities and disperse their members. This makes it difficult for tiny fishes to find regions of adequately concentrated food."

On the April 1978 cruise, with a storm-worthy ship available for the first time, there were no storms. This, itself, becomes the basis for a natural experiment: The calm sea conditions may have fostered the concentration of food species to support optimal growth of post-larval pollack. Were there, in fact, large numbers of young pollack passing through this survival window? The answer will come as the marine biologists follow the 1978 year-class of pollack. If that year-class is large, it may be evidence that calm sea conditions in April tend to foster the survival of pollack; if not, there will be plenty of other data and tentative hypotheses.

In any case, weather is unlikely to be the single most important ingredient of pollack survival. According to Joseph Niebauer, University of Alaska physical oceanographer, in only 3 out of the last 20 years has the Bering Sea shelf been especially calm. Yet the productivity of pollack on the shelf has been good in many years besides those following the calm ones.

Offshoots

A complete explanation of the productivity of the Bering Sea shelf is not yet in hand; there are still two seasons of data collection and analysis ahead. The culmination will be the establishment of a general ecosystem simulation model. Heading this part of the research program are McRoy and John Walsh of Brookhaven National Laboratory. The model will complement a fisheries management model now being assembled by the National Marine Fisheries Service. From the point of view of basic science, the PROBES model should also be applicable, with relatively minor changes, to other Bering Sea shelf species than the pollack.

The power of the model, says McRoy, will be the generality and thoroughness of its explanation of the productive ecosystem of the Bering Sea shelf.

Besides their principal goals—understanding of the Bering Sea ecosystem—McRoy and his colleagues see several ways in which their project may contribute to the theory of ecosystem function in general. They see broad implications in the description of energy transfer upward through trophic levels; on the Bering Sea shelf in the interfrontal zone it appears that the efficiency of energy transfer may be three times as high as was expected. The question arises: Is it likely that this environment alone should possess a uniquely efficient mode of energy transfer? The PROBES study so closely knits the investigation of the several kinds of phenomena bearing on the transfer of energy and materials that "maybe," says McRoy, "we'll be raising the perceived energy transfer level for all oceans." •

The National Science Foundation's support of the research reported in this article has been principally through its Polar Biology and Medicine program.

ACID FROM THE SKY

Acid precipitation knows no national boundaries. The scope of the problem is becoming better known than is what can be done about it.

"The water was so clear it looked transparent. Nothing broke the surface. No frogs croaked. Nothing moved on the shore."

This is what ecologist Anne LaBastille saw not long ago when she hiked in to Brooktrout Lake in a wilderness area of New York's Adirondack Park. Brooktrout Lake had been a prime fishing spot for anglers who didn't mind the 15-kilometer roundtrip hike from the end of the nearest road. But now, LaBastille, a commissioner of the Adirondack Park Agency, wrote in *Outdoor Life*, the trout are gone from Brooktrout Lake. So are other life forms that were linked to the trout and to each other in a complex aquatic ecosystem. The water in the lake, crystal clear with a faint tinge of blue, is so acidic that fish cannot reproduce and survive in it.

Brooktrout is one of the more than 2,000 lakes and ponds in Adirondack Park, 2.4 million hectares of mountains and forests set aside as parkland more than 80 years

Acid-damaged. Leaves on a peony plant grown outdoors in a large city show injury from sulfur dioxide and acid mist.

Environmental Protection Agency

ago. When Carl Schofield, a Cornell University biologist, sampled more than 200 Adirondack lakes in the mid-nineteen-seventies, he found that about half of them were so acidic that fish could no longer live in them.

Schofield and other scientists see convincing evidence that much of the excess acid in the high mountain lakes of the Adirondacks falls from the sky—from rain and snow made acidic by sulfur and nitrogen compounds from power plant and factory smokestacks as far away as Ohio and Indiana. As they travel through the atmosphere and react chemically with water droplets and other airborne substances, these compounds form sulfuric and nitric acid which is then incorporated into raindrops and snowflakes.

Trout fishermen and ecologists are not the only ones who are concerned about acids that fall from the sky. Although the evidence is not conclusive, there are indications that acid precipitation may stunt the growth of forests and reduce agricultural productivity. It certainly is contributing to the deterioration of ancient stone structures like the Parthenon and the Colosseum, as well as other concrete and metal structures all over the industrial world.

Further, the impacts of acid rain have national and international implications. The affected areas may be hundreds or even thousands of kilometers from the sources of the pollutants that form the acids. The United States "exports" ingredients of acid precipitation to Canada; England sends them across the North Sea to Scandinavia; acid haze over Alaska may come from as far away as Japan (see "Alaska's Imported Haze," *Mosaic*, Volume 9, Number 5).

Because emissions from electric power plants appear to be the likeliest major source of acid precipitation, the problem has become a component of energy policy deliberations. "In terms of our domestic energy policy, the acid precipitation problem is pivotal," says Kay Jones, senior adviser for air pollution to the President's Council on Environmental Quality in Washington, D.C.

Although a number of universities and research centers in the United States are studying acid precipitation, Jones says, a carefully planned and closely coordinated national program is badly needed. He has been working with scientists and administrators from other Federal agencies to develop such a program. "I believe we will have a cohesive plan for a national research program on acid rain before the end of 1979," Jones declares.

The focus will be an array of unknowns, including the relationships and interactions

---

How acid is acid rain?

Chemists rate acidity in terms of hydrogen ion concentration, expressed as a factor known as pH, which ranges in value from 0 to 14. A pH of 7 is neutral; values less than 7 indicate increasing acidity, and values higher than 7 indicate increasing alkalinity. The pH scale is logarithmic—a solution with a pH of 5 is ten times as acidic as one with a pH of 6, and a drop in pH from 6 to 4 indicates a hundredfold increase in acidity.

Beer has a pH of 4 to 5; vinegar ranges from 2.4 to 3.4, and lemon juice runs from 2.2 to 2.4. But chemist James Galloway of the University of Virginia warns that pH alone is not a sound basis for comparing acid precipitation with more familiar acids. Lemon juice is a concentrated solution of citric acid, a weak acid, he points out, while acid rain is usually a dilute solution of strong acids such as sulfuric and nitric acid.

Further, rain with a pH of 4 may fall on plant foliage from a sudden summer thunderstorm. When the storm passes and the sun comes out, drops of water standing on leaf surfaces can evaporate rapidly. As the drop grows smaller, the acid becomes more concentrated, possibly to the point that it can physically damage the protective surface structure of the leaf.

Samples of "fossil precipitation" taken from glaciers that formed centuries ago usually have pH values greater than 5. The lower limit of natural acidity for rain or snow is a pH of around 5.6, caused by dissolved atmospheric carbon dioxide that produces a weak solution of carbonic acid. Thus acid precipitation is often defined as rain or snow with a pH of less than 5.6. •

---

among fossil-fuel combustion products, the chemistry of the atmosphere and the acidity of the precipitation that falls from it. The evidence for links between specific pollution sources and particular acid precipitation episodes is mainly circumstantial. The chemistry is still little understood. As ecologist Gene Likens of Cornell University puts it, "The correlations are convincing, like those between cigarette smoking and lung cancer. But I can't prove conclusively that the sulfur in my rain collector in New Hampshire came from a power plant in Ohio."

Forging links

As an example of the lack of knowledge of the mechanisms of acid precipitation, notes Allan Lazrus, a chemist with the National Center for Atmospheric Research and principal investigator for its Acid Precipitation Experiment (see "Research efforts," accompanying this article), more and more nitric acid has been turning up in precipitation samples collected in the northeastern United States, although sulfuric acid has been predominant in the past. It seems probable, says Lazrus, that the reduction in sulfates reflects the trend in fossil-fuel use from high-sulfur to low-sulfur coal and natural gas. Automobile emissions would account for the higher nitrogen-compound levels, he says. But this, he concedes, is an informed guess. It can be neither rigorously supported nor successfully refuted on the basis of present knowledge.

Whatever the precise sources of acid precipitation, it is an increasingly pressing environmental threat. Almost anyplace investigators seek it, it is being found. Recent examinations in the Front Range of the Rocky Mountains by William Lewis of the University of Colorado have uncovered elevated acidity there. And in the Venezuelan tropics acidity levels are being reported "unbelievably" high.

The trends

Although long-term data on trends in acid precipitation over large areas are scarce, it appears that the acidity of rain and snow has probably increased slightly since around 1930. And, according to Cornell's Likens, a pioneer in acid precipitation research, in the last two or three decades "the areas affected by acid precipitation in both Europe and North America have increased rather dramatically."

Recorded observations of acid precipitation date back at least to 1911, when Charles Crowther and Arthur Ruston of the University of Leeds measured a pH of 3.2 (see "How acid is acid rain?" accompanying this article) for rain that fell on that English industrial city that year. In 1939, Henry Houghton of the Massachusetts Institute of Technology, in an early deliberate measurement of acid precipitation, recorded a pH of 5.9 for a rainstorm at Brooklin, Maine. By 1954, Helmut Landsberg, then with the Air Force Cambridge Research Laboratory, was publishing data on rain with a pH of 4.2 from a single storm at Washington, D.C., and an average pH of 4 for rain from 83 storms in the vicinity of Boston in 1952 and 1953.

The first comprehensive precipitation pH data for the United States were collected from a 33-station network by the National Center for Atmospheric Research between 1960 and 1966. Those data showed pH levels ranging from less than 4 in New England to more than 7 in the Western United States.

The seriousness of acid precipitation as a regional problem in the United States was recognized largely as a result of field research done by Likens, Herbert Borman of Yale University and their colleagues at Hubbard Brook

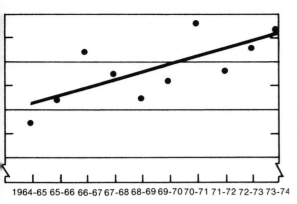
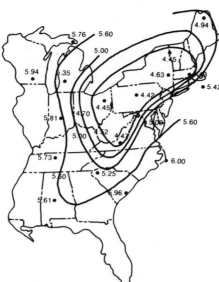
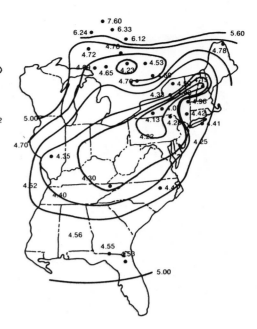

Upward trend. Acidity (scaled at left as hydrogen ion deposition equivalents per hectare) has risen sharply since 1965 at Hubbard Brook, New Hampshire. Increases in the acidity of precipitation over the Eastern United States between 1955-6 (left-hand map) and 1972-3 show in the pH of annual precipitation.

G. E. Likens; Likens and C. V. Cogbill

Experimental Forest in New Hampshire. Precipitation sampling and biogeochemical cycling studies that started at Hubbard Brook in the early nineteen-sixties revealed that rainfall with a pH of less than 4 was a common occurrence there. Other measurements in New England and the Adirondacks, at sites remote from local pollution sources, also showed precipitation pH values around 4. The lowest precipitation pH ever recorded at Hubbard Brook was 2.85, which Likens describes as acidic enough to do serious direct damage to vegetation.

In Europe, systematic monitoring of precipitation chemistry began in Sweden in 1948, was extended throughout Scandinavia between 1952 and 1954 and was expanded to cover most of Europe in the next few years. In 1956, the highest levels of acidity in precipitation were concentrated over southeastern England, the Netherlands and northwestern France and Belgium, with average pH values between 4 and 4.5.

By 1966, the average annual precipitation pH over the Netherlands had dropped to less than 4, and the extent of acid precipitation had increased over much of Europe, as well as Japan and Canada. As in the United States, these increases in acid precipitation accompanied sharp industrial growth and increased energy production.

Although there are still contentions that cause-effect relationships have not been demonstrated, "It's generally accepted," says the University of Virginia's James Galloway, "that precipitation in the Eastern United States has become acidified by fossil-fuel combustion and that this has acidified many lakes in the northeastern United States.

"A few people," he notes, "attribute this to natural mechanisms rather than fossil fuels; the areas of dispute have to do with the amount and quality of the monitoring that's been done."

Causes and effects

Not all lakes that receive acid precipitation become highly acidic. If a lake has a bed of rock and soil that contains alkaline substances, such as the calcium carbonate of limestone, acids may be neutralized as they mingle with lake water.

But in areas where the underlying rock is granite or lava, there is a shortage of such buffering chemicals. Lakes there are acid-vulnerable.

The Canadian Shield, for instance, is a huge formation of granitic rock that extends down across the eastern half of Canada into the Eastern United States. Lakes in this vast, hard-rock region, such as those in the Adirondacks, face a double threat: They are in the path of acid precipitation, and they lack the capacity to neutralize the acid that falls into them.

More than 2.5 million square kilometers of North American land surface overlie such acid-sensitive hard-rock formations. So does most of Scandinavia. Even if a lake has a modest buffering capacity, it can be overridden by large and continuing inputs of acid precipitation.

Likens sees solid evidence that fish cannot reproduce in lakes in which acidity has risen above a certain level. Lakes like those of the Adirondacks seldom grow acid enough to kill mature fish directly, but the acidity can interfere with the more delicate and acid-sensitive reproduction stages of fish.

Acidity can reduce the level of calcium in female fish so low that they cannot produce eggs. Even if eggs form, both the eggs and newly hatched baby fish are vulnerable to comparatively low levels of acidity. Mature fish, subjected to chronic physiological stress by continued exposure to moderately acidic water, can be killed off when the spring snow-melt releases surges of accumulated acid into lakes and streams.

Scandinavia has suffered losses of fish even greater than those of the Adirondack lakes. Trout have disappeared from many Norwegian lakes over the last couple of decades, and whole populations of salmon have vanished from rivers and lakes in both Norway and Sweden. Some went gradually, as acidity slowly increased; others underwent sudden die-offs, when the pH suddenly decreased during periods of snow-melt.

Direct and indirect

The nickel mining center of Sudbury, Ontario, provides an extreme example of the effects of acid on fish. Emissions from the Sudbury smelters, among the world's largest single sources of sulfur pollution, have acidified several hundred lakes within a 50-mile radius to the point at which virtually no fish survive in them.

The link between the nickel smelters and the acidity of the lakes around Sudbury is very direct, of course, and does not depend on complex processes of atmospheric and precipitation chemistry. But the lakes have provided dramatic evidence of the effects of acid on fish for Richard Beamish and his colleagues at Environment Canada, the environmental agency of the Canadian Government.

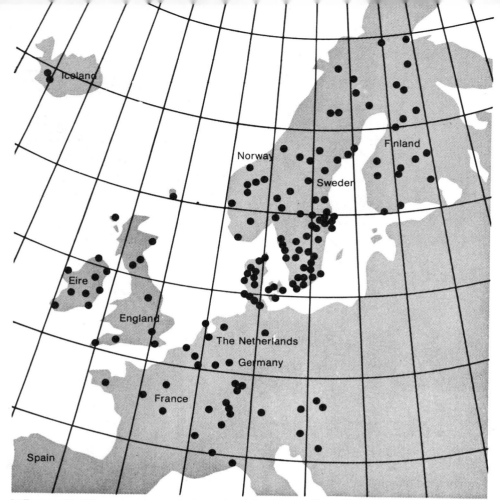

In Europe. The European rain monitoring network (top) is coordinated by the International Meteorological Organization of Sweden. In the Netherlands (below), sulfur dioxide emissions declined sharply in the late nineteen-sixties, while oxides of nitrogen continued to climb.

after A. J. Vermeulen

The first fish to disappear as acidity increased, according to Beamish, were smallmouth bass and walleye. Northern pike and lake trout succumbed next, and even hardy species like lake herring, perch and rock bass finally disappeared as the lakes grew more and more acidic.

The world's tallest smokestack now rises over Sudbury, sending its sulfur compounds out across Ontario and beyond to contribute to acid precipitation hundreds of miles away.

Some indirect effects of acid rain can also have serious impacts on fish. Investigating a drop in the trout population of lakes in which acidity did not seem to be strong enough to account for high fish mortality, Cornell's Carl Schofield discovered a lethal side effect. Strong acid in the rain was releasing aluminum from the soil and carrying it into the lakes.

"We found that the aluminum in this situation is very toxic to fish," Schofield says. The acid runoff with aluminum added was more poisonous to the trout than the acid precipitation alone would have been.

Swedish and Canadian scientists have identified another serious side effect of acid precipitation: elevated levels of mercury in fish

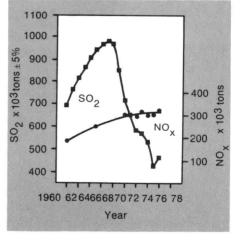

that live in lakes and rivers that are consistently acidic. A Canadian researcher, G. H. Tomlinson, presented a report on this subject to the Panel on Mercury of the U.S. National Research Council in 1977. Tomlinson and his colleagues found that fish in lakes and rivers draining into James and Hudson Bays in northwestern Quebec contained mercury concentrations well above the 0.5 part per million prescribed as the standard for safe human consumption in the United States and Canada.

The mercury apparently reaches the rivers and lakes in very low concentrations in rain and snow and, possibly, by leaching from soil. When the mercury enters well-buffered water having a pH higher than 8, however, it forms dimethyl mercury, a volatile compound that tends to evaporate promptly back into the atmosphere.

But if the pH is 6 or less, highly toxic monomethyl mercury is formed and accumulates in the tissues of fish. Pike in the isolated Broadback River system, for example, often contain more than 2.5 parts per million of mercury, more than five times the level considered safe for human consumption.

Sport fishermen who eat their catch from occasional fishing trips are not endangered by these mercury-contaminated fish. But Indians who eat fish almost every day can absorb dangerous amounts of mercury. In a Canadian Government survey, almost half of more than 700 Indians tested in Quebec, Ontario and the Northwest Territories showed abnormally high levels of mercury in hair and blood samples.

Likens says the question of the effects of acidity on aquatic biological populations other than fish is less certain. But he sees evidence that a variety of other aquatic organisms in the food web may be adversely altered by acid precipitation. In general, fewer or different species of algae and invertebrates are found in acidic lakes and streams. And the rate of organic decomposition is slower. Swedish researchers believe that decreased decomposition of organic matter on lake bottoms, along with increased growth of mosses and fungi, leads to a depletion of nutrients and reduced biological productivitiy in acidic lakes. In short, acidity appears to reduce the ecological diversity and vigor of lakes and streams.

A mix of factors

The effects of acid precipitation on soils and terrestrial vegetation are more difficult to document. "Acid rain is a popular name for something that sounds—or that is—very threatening," says Ellis Cowling of the School of Forest Resources at North Carolina State University, "but it's only one part of a phenomenon that is usually called atmospheric deposition." Cowling points out that many kinds of airborne substances are deposited on vegetation and soils and in surface waters as dry particulate matter, aerosols (tiny liquid particles suspended in air) and gases, as well as in precipitation.

Some of the substances deposited by the atmosphere are injurious, but others, including airborne matter from natural sources, are beneficial. These include spores and pollen from plants, sea spray from oceans, soil particles picked up and transported by the wind, dust from volcanic eruptions and cosmic sources and gaseous compounds of sulfur and nitrogen released by decomposing organic matter.

Cowling maintains that it is important to understand the whole process of atmospheric deposition before attempting to alter any part of it. Atmospheric deposition is a source of plant nutrients as well as damaging substances, he emphasizes, and most plants are nourished through the atmosphere as well as through the soil. "It takes 15 elements to make a plant grow," Cowling says, "and all of them are available in the air. Man is augmenting the supply of both good and bad things in the atmosphere and is changing their rates of flux, but the net effect may not be bad in every situation."

When moderately acid rain falls on a forest, for example, the net effect could be positive because of additional nutrients deposited along with the acid or released more rapidly from the forest-floor humus by the action of the acid. Cowling feels that we need much more detailed and comprehensive knowledge of atmospheric deposition and its effects on forests, grasslands and agricultural systems before we can prescribe effective remedies for the adverse effects of acid precipitation.

Having delivered this *caveat*, Cowling is quite ready to point out that acid precipitation can have many direct and indirect adverse effects on plants. Some potential direct effects are:

• Damage to protective surface structures of foliage.

• Poisoning of plant cells by diffusion of acidic substances into leaves, flowers, twigs and branches.

• Disturbances of normal metabolic or growth processes such as photosynthesis.

• Interference with reproductive processes.

Potential indirect effects may include:

• Leaching of mineral elements and organic substances from foliage.

• Increased susceptibility to drought and other environmental stresses.

• Alteration of host-parasite interactions.

Lab, field and hothouse. Carl L. Schofield studies the survival rate of brook trout against acidified water containing aluminum (top left). Jim Galloway, with John Burke and John Andrews of Woods Hole (top, right), take sediment cores through the ice at Sagimore Lake in the Adirondacks. A multi-year search for acid rain effects on productivity and nutrient cycling in hardwood forests takes place in an enclosure at the Corvallis Environmental Research Laboratory.

Cornell; Environmental Protection Agency

Cowling feels that a simplistic view of the acid precipitation problem may lead to some dangerously oversimplified attempts to deal with it. "One solution that has been suggested," he says, "is to inject ammonia into stack gases to produce ammonium sulfate instead of sulfuric acid. But when plants take up ammonium sulfate, they produce still more acid, which is more harmful than the acid precipitation alone would have been. Instead of solving the acid precipitation problem, you have replaced it with an acidifying precipitation problem."

More than research

Cowling was one of four scientists (the others were James Galloway of the University of Virginia, Eville Gorham of the University of Minnesota and William McFee of Purdue) who recently produced a national plan for research on atmospheric deposition for the President's Council on Environmental Quality. They noted that "reliable research will give no 'quick-and-easy' solutions to the problems of atmospheric deposition, because lakes, forests and agricultural and range ecosystems are complicated communities of organisms. The amount of time required to answer the key questions will vary from months to years. Key questions on the sensitivity of species, lakes and soils to acidification can be answered relatively easily compared to the long-term questions of ecosystem stress and alteration."

If such research efforts, and others that are in progress or being planned (see "Research efforts," accompanying this article), are fruitful, will it be possible to translate their results into effective action to deal with the acid precipitation problem? Or is the whole tangled web of ecology, politics, energy, chemistry, agriculture and other factors too intertwined and sticky to be unraveled? One example of an effort to deal with the problem can be found in the Netherlands, a nation that is much smaller than the United States and much less diverse geographically, politically and economically.

In 1966, the highest acid precipitation measurements in the world were made in the Netherlands. But from 1967 on the acidity of the precipitation decreased, simultaneously with a general reduction in sulfur dioxide emissions from manufacturing and power plants.

One element in the reversal in the Netherlands was effective action by the Government. In 1967, according to Arend Vermeulen of the Netherlands Department of Environmental Control, it was easy to see that a constant increase in energy production, with oil continuing to be the main energy source, would double the emission of sulfur dioxide in five years. "In order to keep the Netherlands habitable," Vermeulen says, "drastic measures were required immediately."

The Dutch Clean Air Act was enacted in 1968 and, in response to its requirements, facilities were built to remove sulfur from industrial emissions, the use of high-sulfur oil was prohibited, and imports of low-sulfur oil were increased.

Total sulfur dioxide emissions dropped dramatically in the Netherlands after 1968, and the acidity of precipitation went down, along with air pollution. But these benefits resulted from luck as well as design and deliberate action. About the time that the clean air act became law, the Dutch discovered the largest coherent natural gas field in the world, and they promptly began to substitute gas for oil. Between 1967 and 1975, the fraction of Dutch energy needs supplied by natural gas rose from 18 to 85 percent, dramatically reducing the emission of sulfur into the atmosphere.

The point of the Dutch experience is not that the United States should hope and pray for natural gas discoveries, but rather that acidity in precipitation seems to be tied closely to patterns of energy production.

As James Galloway puts it, "If scientific research provides good evidence that the ecological impacts of acid precipitation are unacceptable, or that its economic or social costs are excessive, it should be possible for the policy-makers to use that knowledge as a basis for intelligent and effective action." ∎

The National Science Foundation contributes to the research reported in this article through its Atmospheric Chemistry and Ecosystem Studies Programs.

Research efforts

There clearly are many large and critical gaps in knowledge of acid precipitation and its effects. Here are some current research efforts aimed at filling them:

• The Acid Precipitation Experiment (APEX), a cooperative effort by a group of National Center for Atmospheric Research and university scientists, is supported by the National Science Foundation and the Environmental Protection Agency. It includes chemists, ecologists, meteorologists and numerical modelers and is centered on a series of aircraft flights, from Nebraska across Illinois, Ohio, New York, New Hampshire and 150 miles out over the Atlantic. The aircraft crews sample air, cloud-water and precipitation to several levels, while ground-level samples are collected simultaneously at field stations and by a mobile observer with portable equipment. The APEX missions are scheduled to continue into 1981. They are designed to explore and define chemical and physical links between airborne compounds and acid precipitation on a regional basis.

• The National Atmospheric Deposition Program (NADP) was begun by the Department of Agriculture as a cooperative effort by 26 state agricultural experiment staions, with inputs from the U.S. Forest Service, the U.S. Geological Survey and the Department of Energy. It has expanded further and now involves other agencies. Using a network of sampling stations and a central analytical laboratory, NADP is building toward a bank of long-term data on atmospheric deposition over the United States. Its other goals are to determine the relative importance of various forms of atmospheric deposition and to coordinate research on its effects on agricultural lands, forests, rangeland and surface waters.

• Three Adirondack lakes, one neutral, one mildly acidic and one highly acidic, are being studied to try to identify the factors responsible for these three varied responses to acid precipitation.

• The Multistate Atmospheric Power Production Pollution Study (MAP3S) program, begun by the Department of Energy and later transferred to the Environmental Protection Agency, is designed to answer two critical questions: (1) Will precipitation acidity and atmospheric turbidity in the United States increase as coal combustion increases, and (2) can atmospheric concentrations of particulate sulfur be reduced by reducing sulfur dioxide emissions? The program is in three parts, aimed at characterizing the relevant pollutants, studying them through field experiments and developing numerical models of their transport, transformation and effects.

• The Electric Power Research Institute (EPRI) has supported the Sulfate Regional Experiment (SURE), a 39-month study designed to monitor atmospheric pollutants over the northeastern United States and to develop methods for using local emission data to predict pollutant levels in the atmosphere. EPRI has also supported studies of precipitation chemistry and lake acidification.

• Researchers at the Environmental Protection Agency's laboratories in Corvallis, Oregon, have treated plots of maple trees, crops and other plants, sheltered from natural precipitation by a plastic-covered structure, with artificial acid rain. They have studied the effects on yield, growth and quality of this simulated acid precipitation. EPA is currently planning an expanded program of research on the effects of acid precipitation on plants and ecosystems. •

The Ocean in a Test Tube

The resilience of the oceanic food chain is revealed in the Controlled Ecosystem Pollution Experiment (CEPEX).

In the bag. Massive plastic sheaths in which intact segments of the marine ecosystem are captured are inspected before launching.

Three white hoops of steel, each ten meters across, lie like oversized life rings on the breeze-ruffled surface of the quiet water. Scientists, engineers, students and technicians stand elbow to elbow on the rings, waiting expectantly, watching the black-hooded heads of scuba divers bobbing in the cold water.

"Here it comes!" someone shouts. Divers surface inside one of the rings, holding up the ring-wide, open maw of a transparent, double-walled, plastic bag. Workers atop the tubular ring reach down to attach the bag mouth to the ring by thick, rubber straps. They secure the straps to the ring, cursing pinched fingers and broken fingernails, stealing an occasional hurried glance to see how their colleagues are doing on the other rings.

Below the surface, the transparent plastic cylinders extend 23 meters into the darkening greenish gloom. Weighted shroud lines hold the silolike containers vertical against the force of the tidal current. Divers push against the sides to be sure the bags are filled with sea water.

Earlier, the divers had taken the folded cylinders to a depth of 25 meters. On signal they raised the mouths of the unfolding bags to the surface, trapping in each a 40-cubic-meter, vertical column of undisturbed ocean water and all of its natural population of plants and animals. Dacron cones and drain valves closed the bottoms of the bags, giving them the appearance of gigantic test tubes.

Later, in a trailer on shore, David W. Menzel, director of the Skidaway Institute of Oceanography, near Savannah, Georgia, leans back in his chair with a satisfied grin. "We did it," he says, "We cut out a piece of the ocean."

Three intact "plots" of water had been isolated so that Menzel and his colleagues could determine the effects on the encapsulated pieces of ocean ecosystem of a variety of changes to be imposed.

"On land, when researchers want to study the effects of a pollutant or a fertilizer," Menzel observes, "they fence off two plots of ground, add the substance to be studied to one plot and use the other as a control. This has been going on for a long time, but we've just learned how to do this in the ocean. It gives us a new tool in the study of the long-term effects of pollutants on the marine environment."

"The bags can be used as the basis of an early warning system," explains George D. Grice, a biological oceanographer at

Woods Hole Oceanographic Institution. "For example, if you wanted to determine the effect of using a particular site for dumping, a plastic enclosure could be placed at that location first. Materials then would be added to simulate the expected concentration of wastes. The experiment would reveal how the ocean and the food chain handle the stress proposed to be imposed upon it. We also can use the enclosures for a variety of scientific experiments to assess the effects of natural stresses, such as reduced light intensity or nutrients."

Assessing impacts

Assessing artificial and natural changes in the ocean, predicting damaging effects of waste disposal and understanding interactions that lead to the depletion of valuable resources are three major goals of the International Decade of Ocean Exploration (IDOE). A ten-year, multinational effort, IDOE was conceived in the late nineteen-sixties as a means of acquiring the scientific and technical knowledge needed for proper ocean resource utilization and for protection of the marine environment on a global scale. It is a response of the international oceanographic community to the increasing need to turn to the sea to meet the growing world demand for food, minerals and energy. The National Science Foundation has responsibility for the U.S. part of the program, which is divided into four areas: environmental quality, environmental forecasting, sea bed assessment and living resources.

The environmental quality program includes pollution studies that began with a nine-month baseline survey of the oceans in 1971. Follow-on projects involved a geochemical survey of the oceans (see "Circulation of the Oceans," *Mosaic*, Volume 4, Number 3), studies of how pollutants are transferred to and within the world's ocean system and the effects of pollution on marine life forms.

The work so far has led to the conclusion—surprising to many—that the open ocean is relatively free of pollution. "No evidence exists that human activities have raised the general levels of pollutants to a danger point in ocean waters away from coastal areas," declares Robert A. Duce, a University of Rhode Island chemical oceanographer who coordinated the IDOE Pollution Transfer Program. "We have enough accurate measurements repeated by different investigators to feel confident in saying that the oceans away from land are clean—much cleaner than we thought they would be at the start of the IDOE."

This contradicts the image popularized by experiences such as those of explorers Jacques Cousteau and Thor Hyerdahl. For years after their widely publicized voyages—Cousteau in the *Calypso* and Hyerdahl in *Ra*, a copy of an ancient reed boat—it was widely accepted that their reports of oil streaks and balls of tar fouling even the remotest parts of the ocean were true of the oceans in general.

To the allegation: "The oceans are dying!" Menzel counters: "To borrow a phrase from Mark Twain, reports of the oceans' death have been greatly exaggerated."

IDOE investigators had no trouble locating tar balls that came from tank-cleaning wastes pumped overboard by ships at sea. They are concentrated and all too apparent along tanker traffic lanes in such places as the Sargasso Sea, the Gulf of Mexico and off both coasts of Africa. But the South Atlantic and most of the Pacific are clear of them. And where they do occur, IDOE researchers have found,

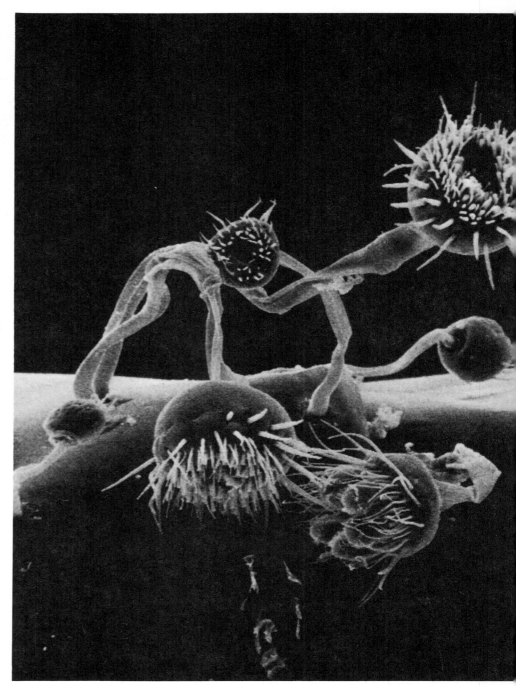

Food chain. A protozoan has colonized the surface of a blue copepod; both inhabit the ocean's surface microlayer, where many ecological interchanges take place.

Test tube. A diver works near the lower end of a controlled-ecosystem enclosure in the Saanich Inlet. Base is closed by a cone through which detritus can be removed.

the harm they do is principally aesthetic. Small animals and plants attach to the balls, and larger animals eat both organisms and tar. There's no evidence that this harms the animals, probably because most of the tar's toxic components have weathered out. Their most serious impact apparently occurs when they wash ashore, fouling beaches and calling dramatic attention to their unsightly presence.

Sublethal impacts

Nevertheless, Duce is careful to point out, and Menzel, Grice and their colleagues agree, none of this means that pollution in the open oceans is not a problem. "We can measure many pollutants at very low levels of concentration in the open ocean," Duce says. "While we know there are no immediate hazards to marine life, we really don't have any idea yet what the long-term, sublethal effects of such concentrations might be."

While specific sources of pollutants, such as sewer and industrial outfalls in coastal areas, chemical runoff and windborne hydrocarbons, continue to be a growing problem, these, the scientists feel, are dramatic and visible and can be monitored and controlled. The specter that haunts them is that of a gradual, undetected buildup of toxic substances in the open ocean. They want to develop a capacity to detect such accumulations before they produce a major catastrophe for marine organisms.

"We are interested in sublethal effects," Grice explains. "We want to determine the concentrations at which metals and petroleum hydrocarbons retard the growth rate of fish, interfere with the reproduction and physiology of copepods on which fish feed, or adversely affect the tiny, floating plant cells upon which all life in the sea ultimately depends."

Plastic bags had been used in the ocean before, for phytoplankton studies. The use of transparent plastic enclosures in which intact columns of ocean water and its natural populations of plants and animals could be trapped and studied for chronic effects, however, originated at a 1972 workshop sponsored by the National Science Foundation. Subsequently, oceanographers deployed quarter-scale (16 meters by 2.5 meters) bags in a pilot experiment in 1974. The following year, full-sized bags were successfully launched. Every year since, IDOE scientists have conducted experiments with both large and small bags in Saanich Inlet, an arm of the Pacific Ocean that penetrates the east coast of Canada's Vancouver Island, near Victoria.

The site was chosen because it combines typical oceanic flora and fauna with sheltered water. In parallel with the U.S.-Canadian work, but on the other side of the Atlantic, the Marine Laboratory of the Department of Agriculture and Fisheries of Scotland began similar bag experiments in 1974. Enclosures 17 meters deep by 3 meters in diameter have been deployed in Loch Ewe, a sheltered Scottish bay populated by oceanic species from the North Atlantic. IDOE scientists call the plastic oceanic "plots" Controlled Experimental Ecosystems, or CEE's; the U.S. IDOE project that utilizes them is called the Controlled Ecosystem Pollution Experiment, or CEPEX.

Resilient populations

Experiments in the enclosures range from two weeks to three months and include organisms from bacteria to fish fry. Work to date has focused on bacteria and plankton, which have been found, as populations, to be remarkably resilient.

Phytoplankton are microscopic, photosynthetic plant cells that are the base of the ocean's food chain. They are grazed by a tremendous variety of zooplankton, minute drifting and weakly swimming animals, fish larvae and the early life stages of virtually all marine animals except mammals. Zooplankton are harvested in turn by larger and stronger species.

Among populations that spend their whole lives as plankton are species that go through one or more complete life cycles in 90 days or less. They make it possible to observe the effects of pollutants in a relatively short experiment. When this was done during the early years of CEPEX, biologists found that, despite massive mortality in the experiments, plankton from microbes to jellyfish and fish larvae recovered from unexpectedly high concentrations of pollutants.

"Mortality sometimes exceeded 50 percent but never reached 100 percent" in CEPEX experiments, comments Menzel, who directed the project from 1972 to 1977. "What's important," he notes, "is the percent that survives, not the percent that dies. As long as a reasonable number of males and females survive, the population eventually recovers," he says.

The shorter the life cycle, the quicker the recovery. "Pollutants affect bacteria first, but they recover in hours to days," says Menzel. "Phytoplankton take three to ten days. Zooplankton require from a few weeks to a few months. Fish would probably need several years; presumably, man, with the longest life cycle, would require the longest time to recover from the effects of pollution."

Vitro vs. vivo

The Saanich Inlet and Loch Ewe experiments provide a much more realistic environment than do laboratory tanks and produce a difference in experimental results as well: Organisms died at lower concentrations of pollutants in the enclosures than they did in the laboratory.

The apparently more realistic enclosure conditions produced a decrease in feeding rates and egg production and an increase in mortality among organisms such as copepods, jellyfish and arrow worms. Those effects were not mirrored in laboratory experiments. So far investigators do not know what causes the difference. They suspect, however, that it involves increased interaction among prey, predators and competitors as well as subtle changes in the amount of organic material in the more natural environments.

"In the laboratory," Menzel notes, "we cannot duplicate the interactions among organisms or the effects of pollutants under the varying chemical and nutrient conditions that occur in the natural environment."

What Menzel and his colleagues have found, however, is that mere numbers do not tell the story; there are apparently qualitative as well as quantitative changes brought on by different stress conditions.

Tailoring food chains

When researchers added very low levels (parts per billion) of mercury, copper or oil to the CEPEX bags, for example, some of the relatively large phytoplankton disappeared. These plants, called diatoms, are ubiquitous in the upper layers of temperate and polar waters and are extremely important to the food economy of the ocean.

With the diatoms gone, smaller (less than ten micrometers), more mobile plants called flagellates replace the diatoms. Such a change—whether brought about by pollution or natural stress—produces a domino effect all the way up a food chain to the fish species valued by man; the kind of stress can, in effect, determine the kind of food chain that will follow.

"If you start with medium-sized plants (30-50 micrometers)," George Grice explains, "the next level in the food chain usually will be large zooplankton. These, in turn, are likely to be associated with commercially important bottom fishes such as cod, halibut and haddock.

"But plants *smaller* than 20 micrometers can be eaten only by small zooplankton,

Experiment in place. Three large enclosures are shown in place in Saanich Inlet (above). A "bongo" net (right) is used to sample plankton in a quarter-sized enclosure in the inlet.

the preferred food of invertebrates such as the comb jellies.

"Still another change emerges if large diatoms (more than 50 micrometers) prevail. They cannot usually be eaten by many zooplankton; fish fry and juveniles consume them. The next level in this food chain would consist of fish caught for fish meal and oil, such as menhaden, anchovy and pilchard.

"If you find an area with large amounts of plant nutrients, such as nitrogen and phosphorus, and microflagellates instead of diatoms, this could mean the area is under stress," Grice explains. "The stress might be pollution, or it might involve natural factors such as light level or water circulation. You'd have to do further work, perhaps with a CEE, to find the cause. But the situation would provide an early warning in commercially important fishing areas like Georges Bank or the North Sea."

CEPEX experiments along these lines reveal that the comb jelly is particularly sensitive to low levels of mercury, copper and oil. Also known as the sea gooseberry, this animal is the size and shape of a large gooseberry; it possesses sticky tentacles,

about ten times the length of its body, and a voracious appetite. Mortality among comb jellies reduces predatory pressure on grazing copepods, leading to a population explosion among these important fish-food species. Results like this suggest the possibility of manipulating an enclosed piece of the ocean for mariculture or fish farming. Diatoms, copepods and other planktonic species might be controlled to produce desired food chains and fish harvests.

No easy answers

CEPEX researchers learned the hard way, however, that the ocean's complexity makes it difficult to generalize about the effects of pollution. "We found a very small difference between the amount of mercury that had no effect and the amount that produced a catastrophic effect," Grice notes. "We saw virtually no effect at one part per billion in the bags. At five parts per billion [about a hundred times that found in normal open ocean water], copepods became drastically reduced. Mercury does not stay at this abnormal level long, however, because most of the metal clings to particulate matter and sinks to the bottom."

The scientists wanted to determine if copepods and other organisms can adapt to a constant input of mercury at low levels, building up resistance that would enable them to survive higher concentrations. In 1977, they kept mercury levels in one large bag at one part per billion by adding metal every two weeks. A second bag received five parts per billion in a single initial dose and then a second dose when levels dropped to about 1.5 parts per billion. A third bag served as a control. The experiment began in late July and ended in October; results are only now being analyzed. CEPEX researchers expect to find an effect on zooplankton; they are not sure, however, whether the metal will be shown to affect the zooplankton directly or to reduce the phytoplankton on which the animals depend for food.

In other, laboratory tests, biologists have found mercury to be three to six times as toxic to zooplankton as is copper. But experiments in the CEE's showed copper to have greater impact than had been anticipated.

Ocean water naturally contains ten times as much copper as there is mercury and, Menzel explains, "Copper in small amounts is a minor nutrient required for plant growth and is present in the blood of many marine animals."

As a consequence, the natural and tolerable level of copper is much closer to a potentially toxic level than is the case for mercury. For example, water might naturally contain one part per billion of mercury and ten parts per billion of copper. If 20 parts per billion of either metal is lethal, pollution would have to increase mercury levels twentyfold to kill organisms, but merely doubling the amount of copper produces mortality.

A difference in toxicity of the same amount of copper at two different locations further confounds the situation. The same concentration of copper produced a greater impact in Saanich Inlet than in Loch Ewe. "There could be several explanations for this," Menzel suggests. But he believes that the correct one involves the difference in organic content of the water. "The toxicity of copper decreases with the amount of organic material available for it to complex with," he explains.

Copper also stays in solution longer than does mercury, so it will be found at greater distances from a pollution source. And, because the mercury attaches itself to particles and sinks more quickly, mercury poses a hazard for organisms that live on the sea bed. Copper, on the other hand, is potentially more dangerous to organisms living higher in the water column.

CEPEX investigators also studied the effects of combinations of metals in the bags. In 1977, they added mercury, copper, lead, vanadium, chromium, nickel, zinc, cadmium, antimony, selenium and arsenic in the same proportions as occur in Narragansett Bay, Rhode Island, an area from which much biological and chemical information is available. By the end of three months, this metal-laced broth took its toll of the young fish and comb jellies living in it. Plankton, however, did not appear to be affected, and, says Gordon Wallace, the Skidaway chemist in charge of the experiments, "These preliminary results appear to indicate that the metals exerted a direct effect on the larger animals."

If levels of metals not high enough to harm plankton can accumulate to poisonous levels in fish, the implications are ominous. CEPEX biologists, consequently, have begun an examination of the fish taken from the enclosure to determine if metals have accumulated in their tissues. "We know that some metals pass right through an animal but others accumulate in their bodies," explains Menzel. "What happens depends on the pollutant, the animal's metabolism and its mode of intake—whether the pollutant is taken in with food or comes directly from the water."

Small bags and oil spills

In addition to work in the big bags, Saanich Inlet experiments have been carried out in the quarter-sized enclosures since 1974. Richard Lee of Skidaway Institute has used the small CEE's to study the effects of oil spills. "A very small percentage of the oil actually gets into the water," he says. "Even below the most gigantic slick, oil concentrations are low—parts per billion." The slick is never solid enough to block the exchange of oxygen and carbon dioxide with the atmosphere, so organisms do not suffocate, as has been suggested. Winds and waves rapidly disperse a slick in the open sea. At the same time, the oil triggers a tremendous increase in bacteria that consume lightweight fractions of the oil, extracting carbon for growth and energy. Phytoplankton take up low-weight fractions not degraded by the microbes and carry them with them to the bottom as they sink at the end of their short lives. Heavy fractions, too large for bacteria to handle, adhere to fecal matter and other detritus and fall to the bottom. "Degradation and sinking occur very quickly," according to Lee.

The oil can produce a decline, however, in the photosynthesis and growth of phytoplankton, with the effects varying widely with species and season. In summer, when nutrient levels are low, for example, plants are more susceptible. But, overall, Lee maintains, "oil pollution is not a potential problem for organisms living in the open sea, with the possible exception of some fish eggs and larvae. Heavy fractions exert their main impact on bottom dwellers living in coastal areas. On balance, it's safe to conclude that the oceans are not dying from petroleum pollution."

Winds and bubbles

"More hazardous than spilled oil," Lee suggests, "may be airborne hydrocarbons produced by burning oil and coal in onshore power plants and by industrial operations such as petroleum refineries. These air pollutants," he says, "along with automobile exhausts, contain polycyclic aromatic hydrocarbons, which produce cancer in mammals. Field work shows these substances to be present in measurable amounts in places such as the Gulf of Mexico. If they become available to marine organisms via their food or the water, polycyclic aromatics could pose a potential hazard."

Researchers on the IDOE Pollution Transfer Program found surprisingly high amounts of petroleum hydrocarbons in the air and discovered how these and other pollutants can be concentrated in surface waters. "One of the major results of the IDOE is the discovery that the top one-tenth millimeter or less is perhaps the most chemically active area of the ocean," says Duce. "This microlayer is enriched in metals, hydrocarbons and sea salts transported from above [by the wind] and below by bubbles.

"Created by the action of winds and waves, bubbles turn out to be an important pollution transfer mechanism," Duce explains. "We once thought the small atmospheric droplets produced when the bubbles broke contained only water and sea salt. Now we know bubbles efficiently transport many pollutants from below the surface and inject them into the air."

CEPEX research has also shown that some organisms, from microbes to fish, possess mechanisms that enable them to handle pollutants that find their way into the ocean. Some marine microbes can volatilize mercury, for example, and laboratory experiments show that the same bacteria that can handle mercury also are resistant to copper and various antibiotics.

"Species that withstand the effects of oil best are the same ones that can best withstand natural stresses such as changes in light and temperature," Lee says. "In other words, organisms that have evolved the ability to survive in stressful environments, such as the tidal zone, are the ones best able to cope with pollution."

The CEPEX experimenters have in no way yet explored the limits of these phenomena. But "nature has an infinite number of solutions to stress," Grice observes, "and we can more easily discover what these are in CEPEX enclosures than we can in the laboratory."

Not all the answers

Although they provide a more realistic environment than does the laboratory, CEE's have drawbacks. The most serious is a lack of horizontal and vertical exchange of water. "In the ocean, organisms and chemicals are transported through a water column by horizontal currents and upwelling," Menzel says. "We have not successfully duplicated these conditions in the CEE's. We are experimenting with upwelling, but we cannot reproduce lateral movements and exchange. In other words, the ocean is so complex we cannot mimic it exactly, but bags have proven to be a better way to deal with the complexity than laboratory or shipboard experiments. We can deal with an entire ecosystem—thousands of organisms and two or three food chain levels—at one time."

In an effort to impose greater realism on the CEE enclosed ecosystems, in 1977 pumps were used to simulate turbulence and vertical mixing in one enclosure. Scientists hoped this would slow the sinking of diatoms to a rate closer to that in the natural environment. The pumps alone did not do this, but by manipulating light levels with baffles and by adding nutrients, experimenters produced a more realistic situation in the enclosure.

It was hoped that the modified light levels and mixing would induce many zooplankton to make the daily journeys that characterize their lives in the ocean. Legions of crustaceans, squid, worms, fish and other animals rise toward the surface after sunset and move back into the depths before sunrise. Light levels in the CEE's may cause many of these creatures to bury themselves in anoxic sediments at the bag bottom in order to find the darkness they seek. The experimental baffles lowered light levels enough to stimulate some daily vertical movement, and more copepods, arrow worms and small shrimp (euphausiids) survived than in the control enclosures. Migrations, nevertheless, did not occur as regularly as they do outside the enclosures, and many animals, particularly the euphausiids and large copepods, smothered in the oxygen-depleted sediments. The only way to prevent this from happening, the experimenters have concluded, is to pump the sediments out of the enclosures at frequent intervals.

Universalities

"These and other experiments conducted in CEPEX have led us to the concept that pollutants produce the same effect as natural stresses have on sea life," Menzel summarizes. "Metals and hydrocarbons, reduced vertical mixing, changes in nutrient levels and alterations in light all seem to trigger the same response. Plant populations switch from relatively large diatoms to small flagellates. Some predators, such as the comb jellies, are particularly sensitive, and the death of these organisms reduces pressure on such grazers as small copepods. These changes govern whether invertebrates, small fish or large fish will dominate higher levels in the food chain. When the stress is removed, the food chain recovers. Exceptions occur, of course, but in general the sequence of events and recovery induced by pollutants or natural stress seems to be universal."

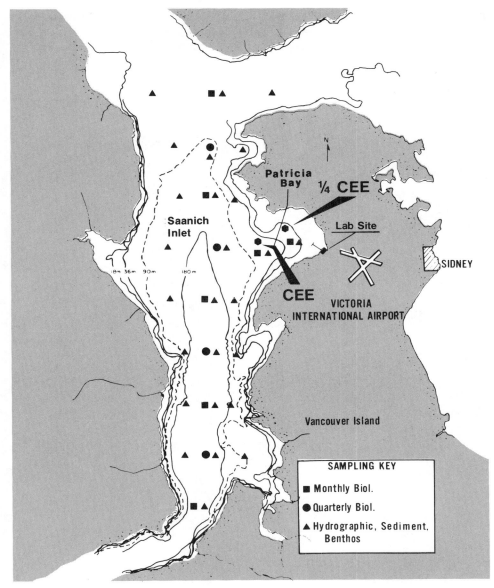

The laboratory. The site of the Controlled Ecosystems Experiment in Saanich Inlet, Vancouver Island, includes locations for both the full-sized and the quarter-sized bag experiments.

CEPEX investigators plan to test this universality theory by experiments in which pollutants will be replaced by natural stresses introduced into the enclosures. Nutrients, light levels and vertical mixing will be manipulated to determine the effect on plant species and populations at the base of the food chain. Predators such as herring larvae, young salmon, comb jellies and arrow worms will be added to the enclosures in subsequent years to evaluate the effects of this kind of stress on lower food chain levels as well as the effects on the predators of variations in the food supply.

"In addition," says Grice, "experiments will be done using pollutants to determine further whether the stresses they impose mimic natural stresses." CEPEX results to date show that low levels of pollutants (one to ten parts per billion) affect all levels of life in the sea. But in many cases, researchers do not know whether decreases in animal populations result from a direct effect on the animal, from an effect on its food, or from an effect on the animals that eat it. Grice, Menzel and other experimenters expect controlled manipulation of natural conditions in the CEE's to resolve the ambiguities.

Marine models

The ultimate goal of CEPEX is to produce enough information for accurate mathematical models of the marine environment to be made. The models would make possible simulations by which effects of man-made or natural stresses on an ocean ecosystem could be postulated. The computerized simulation might conclude that present levels of fishing, pollution or climate change will produce a significant decline in a major fishery. To check this, a CEE could be installed on the site and the environment in it altered to fit that produced by the computer. True effects of this alteration on the food chain—as well as the validity of the model—then could be ascertained.

As an example, scientists point to the North Sea, which is now undergoing three types of regional stress:

- Storm activity has been moving farther south in the past decade; this delays stabilization of the water column that starts the spring bloom of phytoplankton. As a result, the start of the ocean's growing season occurs two weeks to a month later today than it did ten years ago. It is not yet known, however, how this change will affect food chain dynamics in the future.
- At the same time, fishing pressure on the North Sea is growing with the increase in world population and the tendency for nations to exclude foreign fishing vessels from their coastal areas.
- Further, oil drilling and production and tanker operations have increased dramatically in the past ten years.

Can these stresses act synergistically to produce a deleterious effect on one of the world's largest fisheries? If enough data could be obtained from CEE and similar experiments, an accurate simulation model of this complex system could be produced and manipulated. The model's prediction of what will happen if climate change continues and pollution and fishing go unchecked could be verified through additional CEE experiments.

Such experimentally verifiable models would be invaluable tools for those charged with managing pollution or fishing intensity. They would enable the managers to determine the probable sequence of events resulting from their proposed strategies; they would make the establishment of priorities more realistic. Some progress has already been made.

"We are much more optimistic about the ocean now than we were in 1970," Duce concludes. "The general concept of pollution before the IDOE was that it killed everything—that the oceans were dying. IDOE in general and CEPEX in particular have demonstrated that the oceans are a long way from dying; they are in relatively good health."•

Research reported in this article is supported largely by the International Decade of Ocean Exploration program of the National Science Foundation.

Mount St. Helens: The

by John Douglas

The Mount St. Helens eruption became a research opportunit

Virtually moments after Mount St. Helens erupted on May 18, 1980, a massive swath of wilderness in southwest Washington—an area that had once been a lush terrain of mature forest, pristine lakes, and stable streams, a region rich in fish and wildlife—suddenly became an awesome wasteland. Clear lakes became septic broth, fit only to sustain microorganisms. Waterways filled with a dangerously unstable mixture of mud and debris. Plant life in many areas was reduced to a promise of buried root stocks and buried or ferried-in seeds.

The pattern of destruction was roughly wedge-shaped, and covered some 600 square kilometers, the result of an eruption that blew away much of the north side of the mountain.

A lateral blast of gas and debris shot out of the mountain with a force that carried it some ten kilometers before it began to follow the topography of valleys. Propelled by energy equal to that of a ten-megaton bomb, it first sheared off trees within the first few kilometers at about breast height; farther away, it uprooted them. As peaks in front of Mount St. Helens split the blast, the wind of stone, or tephra, left some pockets of vegetation intact while knocking down timber as far away as 20 kilometers. Beyond this, a halo of dead, standing trees stretched out another couple of kilometers in some areas.

Avalanches and mudflows, says Richard J. Janda of the United States Geological Survey, sent slurries sluicing furiously down the South Fork of the Toutle and other rivers. Another flow born of water-soaked avalanche material then moved down the North Fork of the Toutle with such force that it backfilled the South Fork, where the two rivers meet. When things finally settled down, avalanche material lay in front of the mountain to an average depth of 45 meters and a maximum depth of almost 200; mud lay to depths of as much as 15 meters in some valleys and a meter or more across flood plains.

Overhead a plume of ash rose almost 20 kilometers and was carried by winds toward the northeast. Ashfalls heavy enough to coat trees were encountered out to distances of nearly 50 kilometers. The ash stuck in patches to conifer needles, looking like dirty snow and weighing down limbs. An ashfall two and a half centimeters deep was found 400 kilometers away.

Since that major explosion, several smaller eruptions have occurred at Mount St. Helens, but their power has mainly been expended on an area already ravaged. The mountain continued to rumble and threaten into 1981,

Destruction. Rows of flattened Douglas firs, the result of the Mount St. Helens eruption. Dwarfed in comparison are two scientists (circled) and their helicopter (top, right of center). The devastation here was in the blowdown zone about 12 kilometers from the volcano.
Al Levno/U.S. Forest Service.

holding scientists' interest in what had become a landscape of gray desolation: hardened mud swept by occasional floods, fallen trees, a tenacious mantle of rain-hardened ash, and erosion and gullies as winter rain assaulted the denuded slopes.

Ecology of a Holocaust

...precedented magnitude—a chance to see the recovery of a biological community.

Patterns of survival

But there is also life. Soon after the eruption, a pervasive odor of decay coming from lakes betrayed the presence of bacteria and fungi fulfilling their primordial function. A fresh hole on a barren hillside showed where a gopher was trying to reestablish some order in the chaos around him. And in the shelter of fallen trees, plants sprouting from root stock in the old soil tentatively sought the sun. Between the devastation and the evidence of life's tenacity, what had begun as an unmitigated disaster had become an unprecedented research opportunity.

For all its apparent desolation, most of the eruption-affected area around Mount St. Helens never became totally sterile. Rich organic material buried under ash and mud remained a potential support for new vegetation. Throughout the summer of 1980 a variety of small plants began to appear in rapid succession, beginning with those that had survived the blast. Even where all vegetation was killed above ground, plants began sprouting from underground stems (stolons and rhizomes) and roots. Other plants had been under snow when the eruption occurred and later pushed up through the mantle of ash that overlay it. New buds were sent out by some trees that had been badly scorched as well as by others whose older foliage is still covered with ash.

The first task for ecologists was to establish a series of baseline measurements against which to judge the recovery of various life forms. The scale and extent of destruction provided a unique opportunity to witness the succession of species in both aquatic and terrestrial environments. Fundamental theories of ecological equilibrium were and are being tested.

To coordinate the work of the many scientists interested in ecological work around Mount St. Helens and to help them gain access to restricted areas near the mountain, a two-week "pulse" of research was organized in early September, directed by Jerry F. Franklin, James R. Sedell, and Frederick J Swanson, all of Oregon State University. Franklin was in charge of the overall operation and coordinated research activities, Sedell coordinated aquatic work, and Swanson handled the teams studying geological effects and ecosystems, as well as the erosion that raised a clear threat to regeneration in many areas as the winter after the eruption advanced.

The most obvious, costly, and enduring damage was, of course, the destruction of some 600 square kilometers of productive forest—enough timber to provide lumber for 200,000 homes. Douglas firs, western hemlocks, and silver firs, many of them 60 meters tall, were stretched out along the contours of mountains and valleys, tracing the path of the blast. Many others, piled in heaps at the foot of the avalanche, have since been removed by the Corps of Engineers. Close by the volcano, smoke still rose months after the destruction from the smouldering remains of trees. Ignited in the explosion or by lightning induced by the eruption, they were slowly turning to charcoal beneath a mantle of ash.

Delayed effects

The initial eruption came before annual tree growth had started at higher elevations, so many trees, ash-encrusted or scorched

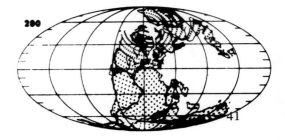

but not killed by the blast, eventually put forth new foliage. Physical damage to needles, temperature stress, and the reduction of photosynthetic potential were sources of initial concern. A team from the University of Washington found that in the heavy ashfall zone old foliage was literally being cooked under a mantle of ash, with leaf temperatures rising to 50 degrees Celsius. Although new growth did occur, it was sharply reduced. The effect, says the University of Washington's Tom Hinkley, may be even more pronounced as the 1981 growing season advances. About 90 percent of the the food that trees needed to support new growth had already been stored by the time of the eruption. But the success of subsequent years' growth will depend on the limited reserves that trees managed to accumulate the first year.

The nutrition problem is being investigated by a University of Washington team led by Charles C. Grier. He and his colleagues have found that roots covered by an ash layer suffered from oxygen deprivation, which restricted water and nutrient uptake. The problem was compounded by the diminished supply of carbohydrates to the roots from the ash-covered foliage. Eventually, Grier says, the plants will come to equilibrium with the oxygen in the soil, and new leaves will appear that do not have ash on them. But in the meantime growth rates will continue to decline and tree mortality will substantially increase—perhaps by as much as threefold— in the affected areas.

The response of the forest understory to the ashfall depended largely on the amount of snow present at the time of the eruption. Donald B. Zobel and Joseph A. Antos of Oregon State University found that, if the ash was not too deep, the presence of snow underneath could help small plants. They were able to emerge through cracks made in the ash mantle as the snow beneath it melted. On the other hand, where snow was deep enough to bury shrubs and small trees, the burden of ash sometimes kept them trapped. Also, deep ash slowed snow melt, reducing plant growth by keeping roots cold and wet. In some instances, says Zobel, there was "spring in September," as June flowers bloomed three months late. But in general, he concludes, "there is a dying back of the woody plants trapped underneath the ash."

Vegetation above timberline on the south side of Mount St. Helens presented a special case, says Lawrence Bliss of the University of Washington. He and his colleagues found that such vegetation, virtually unaffected by the May blast, was also considerably less

Impact. Methane and carbon-dioxide bubbles (top, left) reveal microbial action in North Coldwater Creek Lake. Researchers Marvin Lilley, Cliff Dahm, and James Sedell (top, right) collect samples for gas and nutrient chemistry studies from an 80-degree seep near North Coldwater Creek. Winter erosion (bottom, right) in the Smith Creek basin—an old clear-cut area. Charles Grier, Valerie Winston, and Paul Schulte take soil samples for nutrient-cycling studies.

John Baross; Fred Swanson; John Douglas, by permission.

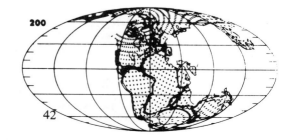

With the "pulse" at Mount St. Helens

As the helicopter descends, a thick cloud of ash gusts, obscuring an awesome vista of flattened timber and melted landscape. For a few moments the pilot jockeys for position, gingerly testing the landing ground for stability. Then he cuts the engine, and an oppressive silence slowly envelops the intruders. No picture of the destruction around Mount St. Helens, no technical discussion of the forces involved, can adequately prepare one for the assault this wilderness makes on the senses.

The touch of each step is strangely unfamiliar; ash compresses unpredictably beneath the feet and a patch of gelatinous mud quivers a warning of quicksand just beyond. From place to place the smell of degradation changes subtly—now acrid and volcanic, now slightly sweet with smoke, now stifling with decay. And everywhere the pervasive stillness and ashen fall remind the wanderer of missing things, such as birds not singing or cool shade lost from the forest floor.

For two weeks in September 1980, more than a hundred scientists representing many diverse disciplines were converging in shifts at the Cispus Environmental Education Center near Randle, Washington, for a chance to study some aspect of the volcano's aftermath. They were participating feverishly in what was being called a pulse of research to establish observation sites and collect what data they could during the short time available. Few surprises were expected to emerge from this first wave of general survey work, but the eventual importance of having many investigators coordinate their research at the site may be considerable. From the base established, a decade-long flow of experiments and papers is likely to develop, tracing the progress of terrain and habitat in a unique environment.

Jerry Franklin of Oregon State University organized the pulse. Surprisingly low-keyed, he not only succeeded in herding dozens of fellow scientists through tightly scheduled helicopter expeditions and logistical maneuvers, but somehow managed to keep peace among the conflicting disciplinary interests and agencies represented. But as an ominous reminder, his voice would drop an octave each time he issued *The Warning* to another departing crew: "The mountain is acting up again; you may have ten minutes to evacuate. Ten minutes!"

The helicopter lands in a small cleared

The "pulse." Norman Anderson, Jerry Franklin, and James Sedell compare notes.
John Douglas, by permission.

space near what is left of upper Clearwater Creek, at the edge of the timer blowdown area some 15 kilometers northeast of Mount St. Helens. Bits of pumice float down the trickling remains of the stream that runs between two lines of hills. Along the farther crest, a surprisingly sharp boundary separates jumbled logs from standing trees, both living and dead—testimony to the thick, rushing cloud of ash and debris the eruption fired down the valley, following the valley's contours.

After a few minutes of awestruck silence and hesitant comments, scientists go about their separate tasks. One patiently searches the seemingly barren streambed of Clearwater Creek for signs of life. Another swings at grasshoppers and damsel flies with his net and terrorizes the leaf beetles on a blackberry bush.

Where Smith Creek, dark and foaming from dissolved organic material, meets the Muddy River, the Muddy's much lighter color reveals its burden of pumice. An odor like that of sewage fills the air. Cautiously picking their way around some quicksand at the edge of the streambed, the researchers drink from a cold, pure artesian spring rising from the debris.

On the way back to the Cispus Center the group flies by the ruptured northern slope of Mount St. Helens, over Spirit Lake and the 180 meters or so of dried mud spread over the upper part of the North Toutle River drainage system. Each new perspective gives a different appreciation of the scope of destruction. From high above, in a commercial jet with passengers crowding to the east-facing windows, one can take a clinically detached view of the eruption. Its scale is evident, but only as a light brown canker upon the velvet green skin of forest. On the ground, all sense of scale is lost amid giant trees tossed helplessly across the line of vision. From a helicopter, one can count the ranges of denuded hills receding into the distance while still picking out, just below, some wrecked car or abandoned machinery that sets a human scale within the destruction.

Each evening during the research pulse, Jerry Franklin would bring the current set of researchers and helicopter crews together to discuss their day's work. Part of the importance of this sort of cooperative research effort is knowing what workers in other fields are doing so that ground sites can be coordinated and mutual frames of reference established. Several joint papers, presenting results from a variety of disciplines, could be anticipated. Yet, even in these informal vesper meetings, the chill hand of research priority curbed discussion somewhat, as researchers who only hours earlier had rejoiced over some new discovery in the field would now shuffle and pass with a brief "Nothing particular."

Nevertheless the conversation of several of the participants in the research pulse reveals a sense of excitement over the prospects for gaining fundamental new insights in their respective disciplines. Beyond the uniqueness of the event itself, several researchers emphasize the importance of the pulse-type approach. As Fred Swanson of Oregon State University puts it: "People have looked at effects of volcanoes in the past; what we can do here is get a much more holistic view. Most results before have represented work by some guy with strong interests. Here we are putting out plots for several decades—even centuries—of study. We want to see long-term processes." Or in the words of Roger Del Moral of the University of Washington—who, like many others, had particular praise for Jerry Franklin's efforts in putting the pulse together—because of this coordinated effort "the whole ecological story at Mount St. Helens should begin to make sense very quickly." •

Support for the biological research "pulse" at Mount St. Helens during the weeks and months immediately following the eruption came chiefly from the Division of Environmental Biology of the National Science Foundation, in special quick-response supplements that enabled grantees to take advantage of the opportunity, and from the United States Forest Service.

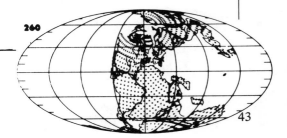

affected by mudflows than that on the northern slope of the mountain. At such an altitude, existing plants were already well adapted to life in raw pumice and severe weather, and most species appeared to be surviving very well.

Survivors included alpine buckwheat, with a deep taproot that allows it to hang on as erosion proceeds, Cascade aster, which has demonstrated a remarkable ability to push up through hardened mudflows, and lupine. The grip of the roots of some lupine species may protect the future mechanical integrity of the soil. Again, scientists are looking particularly to see how root respiration and plant growth are affected by the ground-sealing effects of the eruption.

In the water

Aquatic habitats were also profoundly affected. Forest Service scientists estimate that about 720 kilometers of streams were severely damaged and 60 kilometers completely destroyed by mudflows. Of existing lakes, around 30 were severely damaged, while the damming of streams created many new ponds. Some of these have uncertain futures, since winter rains weakened their retaining structures. As the surrounding forests were blown down, lakes and streams grew dark with dissolved resin and other organic matter. The smell of decay became pervasive.

Even in areas far from the blast site, fish populations were threatened. As streams continued to carry away the ash and mud, they often maintained levels of suspended sediment sufficient to shred the gills of fish within a few hours. Repopulation was inhibited by the loss of some spawning grounds and the deterioration of others. In lakes that still have fish, says James Sedell, who has been studying the aquatic vertebrate populations with Lyle Burmeister of the Gifford Pinchot National Forest, almost all the individuals remaining were adults. "There was essentially no reproduction [the first] year," Sedell says, because blast deposits covering the bottom of spawning areas suffocated the eggs. There was plenty of food for the fish—midges and mosquito larvae quickly returned to the lake surfaces. The problem, Sedell says, was oxygen: The bottoms of many lakes, in some cases to within centimeters of the surface, became anaerobic, and as winter forced the fish to live at lower, anoxic depths, survival rates were sharply affected.

In the view of Robert C. Wissmar of the University of Washington, who with colleagues from the Universities of Washington and Alabama and Oregon State University concentrated on the lakes, a major difference between Mount St. Helens lakes and previously studied volcanically affected water bodies is in the burden of organic matter. Dissolved organic carbon in Spirit Lake increased 48 times, Wissmar reports, while sulfates were up 157 times, and there was a "several thousandfold increase in dissolved iron."

"Counts of total bacteria were so high in many lakes that they could not be accurately assessed," he says. There are, however, a wide variety of specific responses.

The microbial communities in some heavily affected lakes had not yet even begun to get rid of the organic material by midwinter. These systems are chemolithotrophic—dominated by organisms that take energy from dissolved minerals. Such lakes will take decades to recover, though in other lakes aerobic biological processes are already well underway. According to Wissmar, "Given the denuded lake watersheds and continous...erosion, lakes of the blast zone may not recover...until a significant regrowth of terrestrial vegetation occurs."

Ten kilometers or so away from the blast, deposits filling valley bottoms became hot enough to create hot springs in areas that had never had them before. Extremely active nitrification and denitrification could be found going on in the 80-degree-Celsius water in some of these springs, with as many as a billion microorganisms present per milliliter of water.

Initial research concentrated on identifying the important microbial species and biochemical processes present in the the lakes and hot springs. Of particular interest will continue to be the way in which chemolithotrophic processes give way to those involving degradation of organic matter in lakes, and how fixation of nitrogen prepares the way for new species. Within months of the blast, the microbial communities in some small ponds could be seen changing, with colors shifting over a few days. A major task for future research will be to see how quickly the lakes' oxygen levels can be reestablished by the photosynthesis in algae and other aquatic plants. This will largely determine when fish can again be introduced to water that is now uninhabitable.

Mammals

Few animals survived within the devastated region. The Forest Service estimates that perhaps 2,000 deer, 300 elk, and 20 black bears were killed in the blast. But at best these numbers represent guesses based on estimates of initial populations. The main survivors were underground inhabitants like gophers and ants.

So far, research into mammal survival by teams from Utah State University and the University of Florida has found that there has been a reasonable survival of adult gophers, but that virtually no young were present at the end of the summer. It is not yet clear whether this was due to lack of reproduction by females or to starvation of the young; well before winter closed in, the signs of gopher activity within the devastated zone had sharply decreased, indicating possible starvation.

Judging from the number of tracks apparent in the ash even well inside the zone of destruction a few months after the blast, some larger animals were wandering back into the area. Numerous deer and elk were present, but the lack of food and shelter discouraged long visits. In heavy ashfall areas it is still not clear how well large animals escaped the initial blast, but smaller mammals like the chipmunk, red squirrel, and deer mouse appear to have survived.

Soon after the event, the Utah State team spotted a mink at Elk Pass, and coyotes were also seen, indicating that some carnivores are still able to find food. The quantity of ant remains in coyote droppings—coyotes apparently ate the ants for lack of better provision—was evidence of one adaptation. Clearly many years will be required before a normal ecological balance can be established for large animals in the devastated area.

In heavy ashfall areas, by summer, gophers had covered surprisingly large quantities of ground with mounds. At Elk Pass, for example, some twenty kilometers from the volcano in an area of ten to fifteen centimeters of ashfall, a fair amount—up to 2 percent—of the ground was covered with old soil brought up by gophers.

Insects

Insect populations appeared to reestablish through four ways: surviving in refuges, flying in, emerging from eggs, and (in the case of mosquito larvae and maggots) perhaps by being impervious to destruction.

A dozen species of ants were identified early, and together with the gophers they may make a major contribution to preparing new soil by their burrowing activities. Their negative effect on emerging vegetation could also be considerable; one saw, for example, young stalks of bracken fern vigorously attacked by carpenter ants de-

---

Douglas is West Coast representative of Science News.

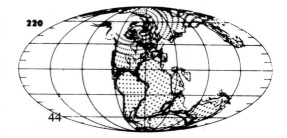

prived of their usual food supply and habitat. The ants were also shepherding and tending the aphids that crowded onto new plant growth within the blast zone or fed in the wounds of standing trees. By late summer, leaf beetles had also begun to feed on emerging shubbery.

Predators of insects, on the other hand, did not fare as well. The eruption apparently caught spiders at a vulnerable time in their life cycle, and birds that glean leaves for insects were probably killed in the holocaust. The researchers will be following carefully the balance between the insects, their food, and their predators over the next couple of years. They may also be able to test some fundamental theories of predation, in particular observing which species of spider return first.

One potentially important effect of ashfall on insects over a wide area was the abrasion caused to species that groom themselves. Ash abrasion of the exoskeleton or damage to the trachea of insects can cause death, and researchers found that bees were particularly hard hit. Because of their importance to pollination—including that of some plants that provide important links in the chain of vegetative succession—a year's loss of bees and other pollinators may hinder plant recovery in some areas.

The long return

Reestablishment of life in the 600 square kilometers of terrain initially devastated by the eruption of Mount St. Helens will depend on many things, including the behavior of the mountain itself. This volcano is still relatively young—most of what was seen before the 1980 eruption was formed within the last 2,500 years—and it has previously demonstrated a wide repertoire of activity ranging from lava flows to explosive eruptions like the most recent one. During a previous active period, it erupted intermittently for more than 15 years; it is still too early to tell how long the current episode may last or what forms the activity may take.

Thriving. Evidence of life returning to the damage zone includes (clockwise, from left) a leopard frog at a new, shallow pond; deer tracks in the ash shortly after the eruption; bracken fern regenerated through the ash from buried rhizomes, within six weeks of the explosion; a gopher hole in a cliff overlooking what was Meta Lake (examined by animal ecologist Larry Harris).

David Johnston; John Douglas, by permission.

Although the ecological transformations dwarf the ordinary human scale, various salvage and restoration activities have been launched. Probably the most urgent are the attempts to mitigate erosion. To some extent, the washing away of loose ash cover is both desirable and unavoidable. In some areas this cleansing action will uncover soil and prepare the way for a new generaion of plants. However, in a climate zone that experiences 170 to 240 centimeters of precipitation a year, concentrated in a few winter months, the impact of uncontrollable floods sweeping masses of debris down river gorges had many scientists frankly worried.

A preliminary Forest Service rehabilitation report warned early that the most critical runoff problems were likely to occur in the heavy blast zone where soil was blown away, reducing water-holding capacity. Farther away, ash deposits over undisturbed soil may have acted to hold soil in place, but loose material could be expected to erode easily on slopes and streambeds. Stream channels were expected to be less stable; lateral meanders could destroy new vegetation. Accelerated snowmelt due to shade removal aggravated the problem of rapid water accumulation. Stream flow was also reduced well below normal, which could add further to the stress on returning plants.

Despite difficult working conditions and arguments among experts over which trees should be removed first, timber salvage operations were launched by major landowners in the region, including Weyerhaeuser and the Washington State Department of Natural Resources. The immediate aims were to protect property downstream and to

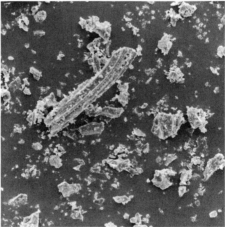

Ashes to... Insects like this honeybee 250 kilometers from Mount St. Helens died agrooming. Abrasive ash particles (enlarged here 450 times) damaged their cuticle layers.

Roger Akre; Allen Crooker, by permission

reestablish productivity of the forest, as well as to recover usable timber.

Weyerhaeuser reportedly expects to be able to salvage about half its downed timber, but the task is difficult and expensive. Logging roads and bridges had to be rebuilt in areas covered by mud and debris many meters deep. Chain saws are dulled by grinding through ash on logs. And although the trees were already on the ground, overall logging costs will end up being higher than normal. Both the industry and the government agencies involved in salvage operations are preparing to plant seedlings to help reestablish the forest.

Another concern of foresters is that bark- and wood-boring beetles and ambrosia beetles were found at many sites containing fallen Douglas fir trees, which would provide excellent brood material. The Forest Service has tentatively concluded that the beetle populations could build during 1981 and then return in massive attacks during the following two years. Whether salvage operations can substantially mitigate that threat remains to be seen.

The rehabilitation of streams is also the subject of controversy. The first priority of the Army Corps of Engineers was to "channelize" major rivers to prevent flooding. The Toutle and Cowlitz Rivers were dredged and leveed to route material quickly downstream, and it is anticipated that the Columbia River will also have to be dredged after winter storms to keep the port of Portland open.

While this plan may prevent some property damage, it could slow the reintroduction of salmon. The Cowlitz system once produced more than 20 percent of the Washington State salmon, but channelized streams make poor spawning grounds. Some spawning may occur in tributaries far upstream, but a Forest Service rehabilitation report concludes that salmon reproduction will be affected for decades.

New symbionts

As research and rehabilitation projects continue, a new symbiosis among scientists from many disciplines is being established around the slopes of Mount St. Helens. By conducting baseline measurements in a coordinated fashion, investigators should be able to construct an unusually complete picture of biological recovery in the region. Several researchers have called the Mount St. Helens event a "once in a lifetime" opportunity. Particularly important will be attempts to see how various systems affect each other—the contribution insects make to vegetation recovery, for example, or the effect of changing soil chemistry on forest regeneration.

Much will depend, of course, on factors far beyond the control of the researchers now planning to devote substantial portions of their careers to the study of the Mount St. Helens area. Further eruptions could suddenly return much of the ecological system back to the starting point again. Winter floods almost certainly made important and unpredictable changes. Full recovery will take more than a century. Indeed, some of the areas around Mount St. Helens had not yet reached their climactic state following eruptions of the mid-1800s. Yet the speed with which many of the biological systems have already begun to reestablish themselves testifies to the resilience of life, as awesome in its way as the destructive power of any volcano. •

The National Science Foundation contributes to the support of biological research on Mount St. Helens principally through its Division of Environmental Biology.

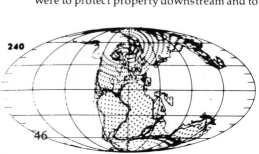

GLOSSARY

acid precipitation. Rain or snow rendered excessively acidic by commercially emitted substances, or substances from other sources (e.g., volcanic activity). These substances react with the water droplets and other airborne materials, producing sulfuric acid and/or other acid compounds. Rain or snow is usually considered "acid" if its pH is less than 5.6.

aerobic. Refers to life or processes that can occur only in the presence of oxygen.

alga (pl. **algae**). Simple one-celled, many-celled, or colonial plants and plant-like organisms often capable of carrying on photosynthesis, although having no true roots, stems, or leaves. Algae are found in water or damp places, and include seaweed, pond scum, diatoms, and many other plant forms.

anaerobic. Refers to life or processes that occur in the absence of oxygen.

Antarctic Convergence. A shifting front or belt at which the antarctic surface water sinks beneath less dense, sub-antarctic surface water. The convergence fluctuates with changing temperatures, pack ice, and currents, moving between 54° and 62° south latitude. The water south of the convergence is more rich and biologically productive than that to the north of the convergence.

arthropod. A member of the phylum *Arthropoda*, characterized by a segmented body and jointed legs. *Arthropoda* contains more species than any other phylum (over 765,000 species), and includes the crustaceans, arachnids, insects, and myriapods.

ash, volcanic. A product of explosive forms of volcanic eruptions, ash consists of very fine particles of shattered rock fragments (called *pyroclastics*). When consolidated, ash forms a fine-grained rock called *tuff*.

atmospheric deposition. Airborne substances that are deposited on vegetation, soils, and surface waters as dry particulate matter, aerosols (liquid particles suspended in air), and gases, as well as in precipitation. Atmospehric deposition can include acid precipitation as well as spores and pollen that are transported by the wind.

bacteria. One-celled microorganisms, usually classified as plants. Bacteria, as a rule, contain no chlorophyll, and therefore are generally incapable of carrying on photosynthesis.

biomass. The total weight (mass) of all living matter in a particular habitat or area.

biota. All species of plants and animals that occur within a certain area.

bloom. A rapid, massive growth of phytoplankton that may occur when large quantities of nutrients, in the proper ratio, are present under conditions favorable for growth.

chemolithotrophic. Refers to a biologic system that is dominated by organisms that take energy from dissolved minerals rather than from the sun (phototrophic).

copepod. One of an order of small crustaceans (see *arthropod*), found in both fresh and salt water, copepods constitute a planktonic species. While some are free-swimming, others live as parasites on fish.

diatom. A microscopic aquatic plant, with over 10,000 species. This alga is most common in cold latitudes and is found in both fresh and salt water. Diatoms are one-celled and are covered with a siliceous, or flinty, outer layer. Often occurring in immense numbers, they play an important part in the economy of nature by accomplishing a major part of all the photosynthesis that takes place in salt and fresh water. They thus serve as a vital source of food materials in many food webs.

ecology. The interrelationships of living things to one another and to their environment. Also, the study of these interrelationships.

ecosystem. The interacting system of biological communities and their non-living environment, in a given area. Most base their food chains on organisms that derive energy from sunlight.

endemic species. A species of plants or animals that is limited in distribution to a small region.

flagellates. Single-celled organisms which move by means of one or more whiplike appendages called flagella. Many flagellates have no cell wall of cellulose, lack chlorophyll, and thus are classified as animals. Others possess distinct plant structures, and are therefore considered to be plants.

food chain. A linear chain of organisms in which each link in the chain feeds on the one before and is eaten by the one after. At the beginning of the chain are the primary producers; at the end are the top predators.

food web. All the interlocking food chains in an ecosystem.

fossil fuels. Coal, oil, and natural gas, so-called because they are derived from the remains of ancient plant and animal life.

fungus (pl. **fungi**). A small, often microscopic plant that consists of single cells or, more often, tubular filaments. Fungi do not contain chlorophyll, and so do not carry on photosynthesis; they are thus dependent on external food producers. There are approximately 30,000 known species of fungi, including, but not limited to, mushrooms, toadstools, molds, and mildew.

genotype. The fundamental make-up of an organism in terms of its hereditary traits; the genetic constitution of an organism.

geochemical. Having to with the chemical composition of the earth's crust.

grazers. Primary consumers (herbivores) that feed on living plants. *See* trophic levels.

hectare. A metric measure of surface equal to 10,100 square meters, or 2.471 acres.

humus. Partly decayed plant matter in the soil.

krill (*Euphausid superba*). A small, shrimplike crustacean. Considered to be a zooplankton species, krill measure about 40 millimeters at adulthood. They feed on smaller zooplankters and phytoplankters and are in turn fed upon by whales, crabeater seals, and other marine life.

microbes. A minute living thing, either plant or animal; a microorganism.

mycorrhiza. A filamentous fungus that forms a bridge between the fallen leaves and the fine roots that cover the floor of tropical forests. The fungus transfers nutrients from the decomposing organic matter to the tree roots, thereby preventing the nutrients from being leached out of the soil by the heavy rainfall.

nitrogen fixation. In nature, the process by which microorganisms and some plants, working together, convert atmospheric nitrogen to a form that the plants can use as a nutrient.

pH. A symbol designating the acidity of an aqueous (water-based) solution. pH is a statement of the concentration of hydrogen ions in the solution, and is represented on a scale extending from 0 (very acid) to 14 (very alkaline), with 7 representing a neutral state.

phytoplankton. The often microscopic plant life found floating or drifting in the ocean or in bodies of fresh water and used as food by higher organisms. Phytoplankton are photosynthetic, and are the base of the ocean's food chain.

plankton. The often microscopic plant (phyto-) or animal (zoo-) life found in bodies of water.

primary producers. Photosynthetic phytoplankton, which on the ladder of tropic dynamics are one rung above raw chemical nutrients. *See* tropic levels.

pumice. A glassy, porous, frothy-looking stone that is formed when certain types of lava congeal after being abruptly discharged through volcanic action. The absence of crystalline structure in pumice is due to rapid cooling, and the frothy character to the sudden release of vapor upon solidification.

rhizome. An underground stem.

stolon. A horizontal stem that produces new plants at its nodes.

tracer species. A species whose growth and development are studied and analyzed with the ultimate goal of gaining a better understanding of the ecosystem of which the species is a part.

trophic levels. The various levels in a food chain, trophic levels in an ecosystem include *producers,* which normally are photosynthetic plants; *primary consumers,* which usually are herbivorous animals that feed on the producers; *secondary consumers,* which feed on the primary consumers; *tertiary consumers,* which feed on the secondary consumers; and *high-order consumers,* which feed chiefly on tertiary consumers. These distinctions are not rigid, as many consumers feed at several different trophic levels.

tropical rain forest. A warm, wet forest in which the temperature rarely drops below 20°C (68°F), and rainfall exceeds evaporation. At least 200 centimeters (80 inches) of rain fall on these forests annually, with some locations getting as much as 1,500 centimeters (600 inches). As the name implies, these forests occur almost exclusively between the Tropic of Cancer and the Tropic of Capricorn.

zooplankton. The often microscopic animal life—including minute animals, fish larvae, and the early life stages of virtually all marine animals except mammals—found floating, drifting, or weakly swimming in the ocean or in bodies of fresh water. Used as food by higher organisms, zooplankton graze on phytoplankton.

INDEX

Acid precipitation
  causes, 29
  direct and indirect effects of, 30-32
  effects upon aquatic populations, 29-30
  effects upon soils and terrestrial vegetation, 30
  impacts, 28-31
  mechanisms of, 28
  recorded observations, 28-29
  sulfur dioxide emissions and, 32
  tied to patterns of energy production, 32
Acid-vulnerable lakes, 29
Acid Precipitation Experiment, the (APEX), 32
Adirondack Park, 27
African Rain Forest, 4
Airborne hydrocarbons, 38
Airborne petroleum hydrocarbons, 35, 38
Alaska Pollack, 20, 26
  larvae, 26
  life history of, 26
Amazon River Basin, 4
American Rain Forest, the, 4
Antarctic Biomass, 11
Antarctic convergence, 11, 13
  seasonal fluctuations of, 11
Antarctic ecosystem, 11, 13
  vulnerability of, 13
Antarctic fur seal, 18
Antarctic marine ecosystem, 13
  exploitation of, 18
  reproductive success in, 16
Asian Rain Forest, 4
Atmospheric deposition, 30, 31

Baleen whale, 11, 13, 16, 18
Bering Sea Shelf, the, 20-25
  extent of, 20
  marine ecosystem, 20
  nutrient dynamics in, 25
  nutrient sampling in, 25
  oceanographic analysis of, 20-25
  primary processes in, 25
  region of interleaving waters, 25
  salinity gradients, 24, 25
  tidal mixing of nutrients, 24
  wind mixing of nutrients, 24
Bering Sea system
  water mixing and nutrient transfer, 24
BIOMASS Program, 18
Brazilian Forest Service, the, 6

Canadian Shield, the, 29
Coachman, L.K., 24, 25
Committee on Research Priorities in Tropical Biology, 2
Controlled Ecosystem Pollution Experiment (CEPEX), 35-39
  goals of, 39

Controlled Experimental
  Ecosystems (CEEs), 35
  37, 38
Cowling, Ellis, 30-32
Crabeater seals, 11, 16, 17
  commercial value, 16
  effects of predators upon
  population, 17
  population regulation
  mechanisms, 16
  scientific observation of, 16

Diatoms, 36
Duce, Robert A., 34, 35, 38
Dutch Clean Air Act, the, 32

El-Sayed, Sayed Z., 13
Euphausia
  antarctic species of, 13
Euphausia Superba, 11, 15
Ewel, John J., 8-10

Filamentous fungi, 7
Franklin, Jerry F., 41, 43

Golley, Frank, 6-9
Golloway, James, 29-32
Grice, George D., 33, 35-37, 39

High tropic level consumers, 22, 23
Hood, D.W., 23, 25
Humid forest ecosystems, 3
Humid tropical forests, 4
Hydrogen ion disposition, 28, 29

IDOE Program, 34
  environmental quality
  program, 34
  Pollution transfer
  program, 34, 38
Indo Malaysian rain forest, 4
International Decade of
  Ocean Exploration
  (IDOE), 34

International Whaling
  Convention, 18
Institute of Ecology, the, 6
Institute of Marine Science,
  the University of Alaska's, 23

Jordan, Carl, 6-8

Krill, antarctic, 11-16
  aggregation patterns, 15, 16
  concentration, 12
  genetic studies of, 15
  genotypes, 15
  harvesting, 11, 13
  larval development, 15
  longevity differentiation of, 13, 15
  population, 11
  swarm classifications, 16
  swarming patterns, 15, 16

Leopard seal, 17
Likens, Gene, 28-30
Loch Ewe Experiment, 35-36

McMurdo station, 16
McRoy, C.P., 21, 23, 25-26
McWinnie, Mary Alice, 13, 15, 16
Medina, Ernesto, 6
Menzel, David W., 33-37
Missouri Botanical Garden, 4, 5
Mount St. Helens, 40-44
  destructive effects, 40, 41
  effects upon aquatic
  habitats, 44
  effects upon insect
  population, 44
  effects upon mammal
  population, 44
  eruption, 40
Multistate Atmospheric
  Power Production
  Pollution Study, the
  (MAP), 35
Mycorrhizae, 6, 8, 9
  role in survival of tropical
  forest, 6

National Academy of
 Sciences (NAS), 2
NAS Committee, 2, 4, 5, 9
National Atmospheric
 Deposition Program
 (NADP), 32
National Center for
 Atmospheric Research, 28
Nutrient-conserving
 mechanisms, 6, 7
Nutrient cycling, 6-8

Oil pollution, 38
Outer Continental Shelf
 Environment Assessment
 Program, 23

Palmer Station, 18
Parmelee, David, 17, 18
Petroleum pollution, 38
pH scale, the, 28
Photosynthetic
 phytoplankters, 21, 25
Phytoplankton, 13, 25, 35, 36, 38
 food source for antarctic krill, 11, 13
 productivity in the Bering Sea, 21
Pollutants, 35-37
 copper, 36, 37
 effects and impacts of, 37
 in coastal areas, 35
 in the open sea, 35
 mercury, 36, 37
 sub-lethal impacts, 35, 36
Prance, Ghillean T., 5, 6, 10
PROBES, 21, 23
 five part program, 23
 research vessels, 23
 theoretical basis, 22
Projecto Flora Amazonica, 5, 6
"pulse", 41, 43

Raven, Peter H., 2, 4, 5, 9, 10
Ross Ice Shelf, 18

Saanich Inlet Experiment, 35, 36, 38
San Carlos Project, the, 6
Schofield, Carl, 28, 30
Siniff, Donald, 16, 17
Skidaway Institute of
 Oceanography, 33
Slash-and-burn agriculture, 7, 8, 10
Sudbury, Ontario, 29, 30
Sulfate Regional
 Experiment, the, 32
Sulfur dioxide emissions, 32

Temperate-zone species
 catalogues of, 3
Tern, antarctic, 18
Timber salvage operations, 46
Toxic monomethyl mercury
 formation of, 30
Tropical forest ecosystems, 2, 4
 destruction of, 2, 4
 role of root mat, 7
Tropical rain forests
 commercial operations in, 10
 preservation of, 9
Tropical reclamation, 8, 9
Tropical species
 catalogue of, 3

Upper trophic-level biomass, 21
U. S. National Marine
 Fisheries Service, the, 23

Venezuelean Institute for
 Scientific Research, 6

Woods Hole Oceanographic
 Institution, 13, 34

Zooplankton, 11, 35-37

574.5
ECO X